JANE GOODALL

Recent Titles in Greenwood Biographies

Jane Addams: A Biography
Robin K. Berson

Rachel Carson: A Biography
Arlene R. Quaratiello

Desmond Tutu: A Biography
Steven D. Gish

Marie Curie: A Biography
Marilyn Bailey Ogilvie

Ralph Nader: A Biography
Patricia Cronin Marcello

Carl Sagan: A Biography
Ray Spangenburg and Kit Moser

Sylvia Plath: A Biography
Connie Ann Kirk

Jesse Jackson: A Biography
Roger Bruns

Franklin D. Roosevelt: A Biography
Jeffery W. Coker

Albert Einstein: A Biography
Alice Calaprice and Trevor Lipscombe

Stephen Hawking: A Biography
Kristine Larsen

Bill Russell: A Biography
Murry R. Nelson

Venus and Serena Williams: A Biography
Jacqueline Edmondson

Flannery O'Connor: A Biography
Melissa Simpson

JANE GOODALL

A Biography

Meg Greene

GREENWOOD BIOGRAPHIES

GREENWOOD PRESS
WESTPORT, CONNECTICUT · LONDON

Library of Congress Cataloging-in-Publication Data

Greene, Meg.
Jane Goodall : a biography / Meg Greene.
p. cm. — (Greenwood biographies, ISSN 1540-4900)
Includes bibliographical references.
ISBN 0–313–33139–1 (alk. paper)
1. Goodall, Jane, 1934– 2. Primatologists—England—Biography.
I. Title. II. Series.
QL31.G58G74 2005
590'.92—dc22 2005016818

British Library Cataloguing in Publication Data is available.

Library of Congress Catalog Card Number: 2005016818
ISBN: 0–313–33139–1
ISSN: 1540–4900

First published in 2005

Greenwood Press, 88 Post Road West, Westport, CT 06881
An imprint of Greenwood Publishing Group, Inc.
www.greenwood.com

Printed in the United States of America

The paper used in this book complies with the Permanent Paper Standard issued by the National Information Standards Organization (Z39.48–1984).

10 9 8 7 6 5 4 3 2 1

CONTENTS

SERIES FOREWORD

In response to high school and public library needs, Greenwood developed this distinguished series of full-length biographies specifically for student use. Prepared by field experts and professionals, these engaging biographies are tailored for high school students who need challenging yet accessible biographies. Ideal for secondary school assignments, the length, format, and subject areas are designed to meet educators' requirements and students' interests.

Greenwood offers an extensive selection of biographies spanning all curriculum-related subject areas including social studies, the sciences, literature and the arts, history and politics, as well as popular culture, and covering public figures and famous personalities from all time periods and backgrounds, both historic and contemporary, who have made an impact on American and/or world culture. Greenwood biographies were chosen based on comprehensive feedback from librarians and educators. Consideration was given to both curriculum relevance and inherent interest. The result is an intriguing mix of the well known and the unexpected, the saints and sinners from long-ago history and contemporary pop culture. Readers will find a wide array of subject choices from fascinating crime figures like Al Capone to inspiring pioneers like Margaret Mead, from the greatest minds of our time like Stephen Hawking to the most amazing success stories of our day like J.K. Rowling.

While the emphasis is on fact, not glorification, the books are meant to be fun to read. Each volume provides in-depth information about the subject's life from birth through childhood, the teen years, and adulthood. A thorough account relates family background and education, traces personal and professional influences, and explores struggles, accomplishments, and contributions. A timeline highlights the most significant life events against a historical perspective. Bibliographies supplement the reference value of each volume.

PREFACE

Born in 1181 or 1182, Giovanni Francesco di Pietro Bernardone was the son of a prosperous merchant. About 1205, while Giovanni was still in his early twenties, he renounced his inheritance and adopted an austere life of poverty and faith. Caring for the sick, the infirm, the old, and the dying, he became ever after known as Francis of Assisi.

In her work with the chimpanzees of Gombe, and in her general devotion to animals and the environment, Jane Goodall has operated within the tradition that Francis established at the beginning of the thirteenth century. Besides his recognition of human fellowship, Francis also regarded the whole of nature as his "brother" and his "sister." He was the first writer to emphasize the beauty and goodness of creation, and to introduce into Western thought the idea that human beings had an obligation to care not only for each other but for all living things.

During the centuries that followed Francis's death in 1226, the human respect for, and stewardship of, nature took many forms. Nearly 800 years later, Jane Goodall has continued to nurture and advance that inheritance in her efforts to bridge the human and animal worlds. It has been an extraordinary and rewarding enterprise.

From the outset of her career, Goodall never distanced herself from the subjects of her study. She sought not merely to attain objective, scientific knowledge, but instead pursued a deeper understanding of chimpanzees and their relation to human beings. Goodall sensed that the kind of understanding she wanted to achieve could arise only through empathy with the chimps and participation in their lives. This modest approach enabled Goodall to avoid imposing her own preconceptions on them. Instead, she allowed the chimps to tell her about themselves. Her methods, which as often as not brought censure from the scientific community, proved the source of her most groundbreaking insights. Goodall's research and fieldwork showed that chimpanzees could think, feel, and communicate in ways that approximated human capacities.

In addition to revolutionizing the way scientists viewed chimpanzees, Goodall has also felt a responsibility to care for and protect them. In recent years, she has done her utmost to acquaint governments around the world and the public at large with practices and conditions that threaten the chimps' welfare and existence. Although she accepts the use of chimpanzees in scientific and medical research, for example, she has pressed for the development of computer models and the extraction of tissue samples that would enable research to go on without chimps. She has also criticized the treatment that chimps receive in laboratories and other research facilities. At the same time, she condemns poaching and the capture of chimps for zoos and circuses, activities motivated by greed and ignorance that have dramatically reduced the chimpanzee population. "We must speak for them," Goodall insists, "for they cannot speak for themselves."

Over the years I have had the opportunity to write about the lives of many interesting men and women from different periods in history and different walks of life. I completed each project feeling that I had unlocked some of the secrets of lives well lived. With every biography I wrote, I gained new insight into what makes great people great. While vision, heroism, and courage are aspects of any successful life, other qualities stand out even more fully, including decency, compassion, empathy, respect, hope, and faith. Researching and writing about the life of Jane Goodall has differed from my previous experiences. My study revealed not only a life well lived, but touched something deep and personal inside me. Jane Goodall has now dedicated more than 40 years of her life to speaking for and with chimpanzees. I marveled at her commitment to, and was moved by her love for, the chimpanzees at Gombe. I admire her quest to make the world a little less barbarous and a little more humane. For through the years, Goodall's mission has remained constant: to remind human beings of their unique obligations toward all the creatures with whom they share the earth.

INTRODUCTION

In recent polls, Jane Goodall has emerged as the most easily recognizable living scientist in the Western world. Her work with the chimpanzees at Gombe has been renowned as one of the great achievements of scientific research. Goodall's research and dedication are held in the same regard as Albert Einstein's. Her approach to field study, once ridiculed and challenged by the scientific world, has now become the model for other ethologists to use. Young girls want to be like her; adults respect her. Children across the world have tried to help her. Throughout it all, Goodall stays steady as she continues to bring her message of peace, hope, and challenge to make the world—for all living creatures—a better place.

Jane Goodall's childhood dream to live among, study, and write about wild animals in Africa seemed the stuff of which fantasies are made. Even now, it takes little effort to imagine Goodall sitting at the big oak dining room table in a warm and comfortable English home, explaining to her grandchildren what she wanted to do when she grew up. It is much harder to conceive of this woman, still slender and attractive at the age of 70, recounting all that she has accomplished during more than 40 years studying chimpanzees at the Gombe National Park in Tanzania, East Africa. The improbability of her story only deepens when the many obstacles she encountered and overcame come more clearly into view.

Goodall was a pretty, demure, though somewhat flirtatious, woman in her early 20s when she accepted the invitation of a friend to visit Africa in 1957. Because her family could not afford to send her to a university, she did not have a college degree. Yet, she was well read, thoughtful, intelligent, and determined. These qualities favorably impressed the noted but controversial anthropologist Louis Leakey, then curator of the Museum of Natural History in Nairobi, Kenya, who offered Goodall the opportunity to research chimpanzees in their native habitat.

With a vested interest in preserving the status quo, the scientific community demeaned the project and doubted the qualifications and abilities of Leakey's untutored protégé. As is often the case, the experts were wrong. Goodall's unconventional approach to her work yielded stunning results. Contrary to accepted scientific opinion, which dismissed chimpanzees as uniformly brutish and vicious, Goodall found chimps to be capable of a wide range of emotions, including affection, compassion, and love. Goodall also showed that chimps could reason, think, and solve problems. They formed hunting parties to capture small monkeys, baby baboons, and baby bush pigs. Scientists had previously assumed that chimpanzees were vegetarian. More startling, Goodall discovered that chimpanzees could fashion primitive implements from grass, twigs, and leaves, dispelling the notion that humans were the only species that made tools.

Goodall drew these and other conclusions about the behavior of chimpanzees only after months of careful, painstaking observation. Yet, contrary to the accepted standards and procedures of field research, she had also encouraged social interaction with the chimps in an effort to win their trust. The emotional attachment that Goodall developed with her subjects, symbolized by her unorthodox practice of naming them rather than identifying them by numbers or letters, prompted many scientists to question the validity of her findings. Her attribution of human characteristics to chimpanzees, her critics have charged, robbed Goodall of the critical distance essential to making objective judgments.

Goodall not only broke new ground in research and field methods, she also opened the door for many others, notably Dian Fossey and her work with gorillas and Biruté Galdikas and her studies of the baboon. Goodall has never considered herself a feminist, nor was her desire to join the ranks of other noted ethologists and in doing so breaking a gender barrier. All she ever wanted to do was observe her chimps. But through her work, Goodall brought a woman's touch, a view that emphasized relationships rather than rules, to be receptive rather than controlling, to be empathetic instead of objective. Her approach flew in the face of conventional science, a science dominated and defined by male views and values. It was a startling break and one that earned Goodall a reputation as a maverick.

When Goodall first began observing the chimps at Gombe, they were little more than black blurs at the end of her binocular lenses. Their movements appeared jerky as if in rapid motion like an old black and white film. As the days and months passed, Goodall began seeing the chimps in a new light; as personalities with likes and dislikes, who had emotions much like humans and who were capable of noble deeds and base acts. More important, she saw the beginnings of what defined humans, their early heritage and the journey traveled to become straight-standing men, women, and children.

In the first 18 months of her study, Goodall amassed over 850 pages of observations, with nary a measurement or number in sight. Instead she wrote, describing in words, devoid of cold technical terms, what she saw and felt. She had no theories to prove; she was content to watch as the chimps slowly invited her into their world. She never took for granted the gift she had been offered by these remarkable animals. All she could do was to present, as accurately and as passionately as possible, the amazing discoveries she had witnessed at Gombe.

Goodall brought to her work an intuitive rather than deductive or objective viewpoint. She named her chimps instead of assigning them numbers. She stepped in to do what needed to be done in order to help chimps who were ailing, hurt, or dying. At a time when the field of ethnology was becoming more technical, impersonal, and statistical, Goodall implemented a new approach that emphasized the feminine point of view. She made friends with certain chimps. She laughed at their antics, appreciated their feelings, was happy when they gave birth. She wept at their deaths. Without them, she would never have had the remarkable journey that she has traveled over the last four decades.

Goodall has suffered through trials in her life: a divorce, the death of her second husband, a deep spiritual crisis, the criticisms, and the naysayers who find her current mission unduly optimistic. Through it all, Goodall has remained steadfast. She is convinced that objectivity in science, or any intellectual endeavor for that matter, is not only impossible but also undesirable. Although researchers must be thorough and honest, they cannot and should not set their emotions aside, because understanding is always personal and participatory. That humility may have compromised Goodall's objectivity, but at the same time it has also made her more aware of, and more sensitive to, the integrity of her subjects and the potential harm that researchers might do to them. She is convinced that the world, its lands, waters, and air can be saved and that animals will finally receive the respect they so richly deserve.

Today, Goodall's work continues through the Jane Goodall Institute for Wildlife Research, Education, and Conservation, though Goodall herself can spend no more than a few weeks a year at Gombe. She has instead expanded the focus of her concerns, traveling and lecturing about chimpanzees, primate research, environmental conservation, and world peace. Yet, time has vindicated Goodall's methods as much as it has certified her most important discoveries. She is now, and has long been, recognized as the principal authority on chimpanzees in the world. It is the improbable fulfillment of a little girl's dream.

TIMELINE

1934	Jane Goodall is born April 3 in London England
1939	Goodalls move to Bournemouth
1942	Vanne and Mortimer Morris-Goodall divorce
1952	Goodall moves to London to begin secretarial school and find work
1957	Takes first trip to Africa
	Begins work with Louis Leakey at Olduvai Gorge
1958	Leakey asks Goodall to take on a field study of chimpanzees
1960	July—Goodall begins her field study at Gombe Stream Game Reserve
	October—Goodall observes David Greybeard fishing for termites
	November—Goodall sees chimps eating meat
1961	Goodall receives her first grant of $1,400 from the National Geographic Society
	Summer—David Greybeard's first visit to camp
1962	Goodall enters Cambridge as a graduate student to begin work on her Ph.D.
1963	Goodall meets wildlife photographer Hugo Van Lawick
1963	Goodall receives the Franklin Burr Award for her contribution to science
	Goodall's first article, "My Life Among the Chimpanzees" appears in the August issue of *National Geographic*
1964	Goodall marries Van Lawick, March 28

	Flo gives birth to Flint, which allows Goodall to observe chimpanzee parenting
	Mike gains dominance in the chimpanzee community
	Gombe Stream Research Center in founded, December
1965	Goodall receives her Ph.D. in Ethology from Cambridge University, becoming the eighth person to ever receive the degree without a Bachelor's Degree
	Second article, "New Discoveries Among Africa's Chimpanzees" appears in *National Geographic*
	Film "Miss Goodall and the Wild Chimpanzees" is broadcast on American television
1966	Fifteen Kasekela chimps are afflicted with polio
1967	Hugo Eric Louis Van Lawick (Grub) is born
1968	Hilali Matama is hired as Goodall's first official field assistant
1970	Kasakela-Kahama (KK) community began to divide.
1970–1975	Goodall appointed to faculty at Stanford University
1971	*Shadow of Man*, Goodall's first book, is published
1972	Goodall's first children's book, *Grub, the Bush Baby*, is published
	Goodall's mentor Louis Leakey dies
1974–1977	Kasakela community males begin a series of attacks against members of the Kahama community that led to the deaths of all Kahama males. This is the first record of "warfare" in nonhuman primates
	Goodall and Van Lawick divorce
1975	Goodall marries her second husband, the Hon. Derek Bryceson
	Four researchers kidnapped by Zairian rebels; although they are unharmed, outside researchers are not allowed into Gombe until 1989
1977	The Jane Goodall Institute for Wildlife Research, Education, and Conservation is established in San Francisco, California
1979	Goodall publishes article "Life and Death at Gombe" in *National Geographic*
1980	Goodall receives the Order of the Golden Ark, the World Wildlife Award for Conservation presented by Prince Bernhard of the Netherlands
1984	Goodall's second television special, "Among the Wild Chimpanzees" airs

	The Chimpanzoo project is created to continue research on chimp behavior while promoting more stimulating environments for chimps and other captive primates
1986	*The Chimpanzees of Gombe* is published
1987	Goodall receives the Albert Schweitzer Award from the Animal Welfare Institute in Washington, D.C.
1988	Jane Goodall Institute-UK is established in London
1988	Goodall receives the Centennial Award from the National Geographic Society
1990	Jane Goodall Institute-Tanzania is launched in conjunction with Goodall's 30 year anniversary at Gombe
	"Chimps Like Us" an HBO documentary airs; the program is later nominated for an Academy Award
	Through a Window is published
	Goodall receives the Kyoto Prize for Science, the Japanese equivalent of the Nobel Prize
1991	Roots & Shoots program is created
1992	Tchimpounga Sanctuary for Chimpanzees is created in the Congo Republic
1993	First video of a chimpanzee birth is recorded
	Visions of Calaban: On Chimpanzees and People is published
1995	Kitwe Point Sanctuary for Chimpanzees established in Tanzania
	Goodall receives the National Geographic Society's Hubbard Award for Distinction in Exploration, Discovery, and Research
	Goodall is awarded the title CBE (Commander of the British Empire) by Queen Elizabeth II
1996	Goodall receives the Tanzanian Kilimanjaro Medal for her work in wildlife conservation
1998	Jane Goodall Institute-Uganda Sanctuary relocates 19 orphaned chimps from other overcrowded sanctuaries
1999	*Reason for Hope: A Spiritual Journey* is published
	Hugo Van Lawick dies
2001	Vanne Goodall dies
2004	Goodall invested as Dame of the British Empire by Prince Charles for her service to the environment and conservation
	Goodall receives the 2004 Nierenberg Prize for Science in the Public Interest

Chapter 1

A CHILD OF DESTINY

If early interests in a young life offer a clue to a child's destiny, then Jane Goodall forged her path at a tender age. The first daughter of Mortimer Herbert Morris-Goodall and Margaret Myfanwe Joseph, Valerie Jane Morris-Goodall was born on April 3, 1934 in a London hospital. Four years later, the Morris-Goodalls welcomed a second daughter, Judy, into the family. Later it was discovered that the two girls suffered from a neurological condition known as *prosopagnosia,* or memory impairment for faces and patterns. Goodall could not recognize people's faces, more interesting is that she had no trouble recognizing animals, especially the chimpanzees that she would one day meet.

During the 1930s England was relatively quiet. The First World War, which had robbed the country and Europe of their innocence and belief in inevitable progress, had ended more than a decade before Jane's birth. Although England, like much of the rest of the world, suffered in the throes of depression, and although menacing governments had come to power in Russia, Italy, and especially in Germany, the Goodalls' world was tranquil.

A WORLD OF CONTRASTS

Jane Goodall's early life was, however, marked by dramatic contrast. Her father, who had once worked as a telephone cable testing engineer in London, had, by the time of Jane's birth, discovered his love of racing cars. Herbert Morris-Goodall's passion for speed was such that he left his job to become a racecar driver. As a member of the British Aston-Martin race team, Morris-Goodall's life was transformed into an endless round of competition. Traveling from track to track throughout England and the European continent, Morris-Goodall soon earned a reputation for being a skilled driver as well as a gentleman whom other drivers held in high esteem. His career spanned more than two decades, and by the time of his retirement in the 1950s, Morris-Goodall had not only earned national glory, but also the distinction of being the only British driver to have competed 10 times in the famous and grueling Le Mans Grand Prix.

The world that Morris-Goodall inhabited—a world of noise, speed, and danger—did not impress his eldest daughter, who showed little interest in her father's passion. Her mother, "Vanne," claimed Jane's time and lovingly kindled the spark that in time guided her to Africa.

Vanne introduced Jane to nature, beginning with the family's backyard garden at their house in suburban Weybridge. They also visited the many public gardens in London. Jane's earliest memories are of wandering with her mother in the backyard. Earth, sunlight, birds, bees, insects, and other small creatures caught Jane's eye and fired her mind and imagination. She recalled that: "I was lucky enough to be provided with a mother wise enough to nurture and encourage my love of living things and my passion for knowledge."[1] From these garden wanderings developed Goodall's intense passion for animals.

One of Vanne's favorite stories about Jane involved Jane's discovery of a bed of earthworms. One day, Goodall's nanny ran from her young charge's bedroom to Vanne, exclaiming that Goodall had hidden in her bed a small handful "horrible, pink, wriggling worms."[2] The nanny also explained that the worms had been hidden under Jane's pillow and she adamantly refused to let them go. Vanne later recalled: "A peach-colored light from the setting sun was flooding the nursery. Jane's eyes were already closing, one hand was out of sight beneath the pillow. I pointed out that the little creatures would find it altogether too hot and stuffy under the pillow."[3] Rather than getting angry about the mess Jane had made, her mother instead explained: "Jane, if you keep them here they'll die. They need the earth."[4] Although she was only 18 months old, Jane understood and obeyed, returning the worms to their garden home. Vanne later admitted that the entire incident meant more to Jane than the simple act of having to give up her newfound "pets." But in later years, Vanne believed that Goodall's interest in the natural world was so intense that her curiosity about nature and desire to interact with it were more mature and deep than would be expected from a young child.

Vanne Morris-Goodall's greatest desire was to provide a secure and stable life for her family, though she also wished to expose her daughters to the harsh realities of the world. As a result, Jane learned early that man and nature did not always coexist peacefully. After watching a man crush a dragonfly that was hovering about her, young Jane cried, "not because I was afraid but because I felt bad that such a pretty thing was destroyed."[5]

In 1936, when Jane was two years old, London residents were enthralled by the first birth of a chimpanzee ever to take place at the London zoo. The new chimp, named Jubilee, became an instant local celebrity. In honor of the new arrival, an English toy manufacturer produced a series of stuffed "Jubilee" chimps for young children. Even Morris-Goodall was caught up in the excitement and purchased one of the toys for Jane.

For Goodall, the toy was love at first sight. Although many of her mother's lady friends were aghast, believing that such a large and ungainly toy would frighten Jane, Jane herself clutched the animal tightly to her chest. The stuffed chimp became her most prized possession, accompanying her wherever she went. It was the first indication of what would become her life's work.

In 1939, when Jane was five years old, her father moved his family to France to be nearer the major racing centers. He also wanted his daughters to learn the language and be educated there. But his plans soon went awry, for only months after the Goodall's arrival in France, on September 1, 1939, the Nazis invaded Poland and began the Second World War in Europe. By June 1940, Germany had conquered France. When he learned what was taking place in Eastern Europe, Morris-Goodall did not hesitate. He immediately moved his family back to England, settling them in the home of his paternal grandmother in the Kent countryside.

Goodall adored her new surroundings. Sheep and cattle grazed and lulled near the old manor house, which was built of gray stone. Not far from the house were the ruins of an old castle where King Henry VIII had imprisoned one of his six wives, but which now housed families of spiders and bats. Her grandmother's house was still lit by oil lamps instead of by electricity. But for Goodall, that made the house all the more magical. At the very moment that Europe descended into war, Jane entered one of the most idyllic periods of her life.

Jane's interest in nature continued to grow and deepen. On a visit to her grandmother's farm, she was asked to collect eggs. She was more than happy to oblige, but soon became annoyed that she did not understand how the hens formed and laid the eggs, telling an interviewer years later, "I had always wondered where on a hen was an opening big enough for the egg to come out."[6] When the explanations she received did not satisfy her, Jane decided to find out for herself.

Following a hen into the henhouse, Jane resolved to sit and watch for as long as it took to learn the secret. But after the hen, upset at her presence, began squawking, she decided to wait for the hens to leave the chicken house so that she could enter it unobserved and conceal herself inside. Crouching quietly in a corner, she waited until a hen entered and settled into its nest. Soon the hen stood up and Jane noticed a small white egg emerging from between its legs. With mounting fascination, Jane watched as the hen shook its feathers, nudged the egg out, and left.

Later that afternoon, Vanne returned home, only find the house quiet and empty. She found out that everyone had gone in search of Jane, who had by now been missing for more than four hours. By seven o'clock concern had been replaced by worry from her family and neighbors. Vanne later wrote about Jane's homecoming: "I don't remember who saw her first—a small disheveled figure coming a little wearily over the tussocky field by the hen houses. There were little bits of straw in her hair and on her clothes but her eyes, dark ringed with fatigue, were shining."[7] When Jane at last appeared, her mother was more relieved than angry. Sensing Jane's excitement, Vanne Goodall did not scold her but instead listened to the story of her adventure. Jane then exclaimed, "So now I know how a hen lays an egg."[8] Vanne, looking at her weary and disheveled daughter, took her hand and walked her home. At the same time, Vanne recognized that her daughter had achieved a certain success with her visit to the hen house: Jane had completed her first animal field study.

A WORLD AT WAR, A LIFE AT PEACE

Jane's idyllic life did not last long. On September 3, 1939, two days after the Nazis had invaded Poland, the government of Great Britain declared war on Germany. Years later, Jane's memories of that day remained vivid. Gathered in the drawing room, the family heard the sad announcement on the radio. For several moments, the room fell silent. As young as she was, Jane knew something had changed and had a sense that it was terrible. Like most Englishmen, Herbert Morris-Goodall resolved to fight. He wasted little time enlisting in the British Army and was assigned to the Royal Engineers, serving first in Europe and later in the Pacific.

With her husband gone, Vanne decided to leave Kent and take her two daughters to stay with her mother, "Danny Nutt" as Jane called her, at the Birches, a lovely red-brick Victorian home located in Bournemouth not far from the English Channel. Also staying at the home were Vanne's two sisters. In this all female household Jane spent the rest of her childhood. She continues to call the Birches home whenever she returns to England. One of the few men to come to the Birches during this time was Vanne's older brother, a surgeon in London, who visited every weekend. Not long after Vanne and her daughters came to live with her, Jane's grandmother opened the Birches to women left homeless by the German bombardment of London.

Despite the upheaval that the war had brought, Jane found life at the Birches to be pleasant. There was so much to see and do; everyday promised a new adventure for a spirited girl with a fertile imagination. The Birches had a large garden in which to roam, and there were countless bushes and trees just waiting to be used for hiding and climbing. Jane spent as much time outside as she could, watching the birds make their nests, the squirrels scurrying to and fro and gathering nuts, and the spiders spinning their webs. At the Birches, Jane also enjoyed the company of several cats, guinea pigs, a number of hamsters, a few turtles, and a canary named Peter. Another favorite pastime in which Jane and her sister Judy engaged was racing snails on whose shells the girls had painted small numbers.

But it was a dog named Rusty, a black mongrel with a white patch on his chest, who had the most profound influence on Goodall. "The Little Black Man" as Jane affectionately called him accompanied Jane on many of her adventures. Rusty belonged to the owners of a hotel around the corner. Each morning at six, he would arrive, bark outside the door until Jane came for him, and stay with her all day except when she was in school. Jane had trained him to do tricks such as closing a door and rolling over. Many years later, Goodall stated that if hadn't been for Rusty, she wouldn't have been nearly as effective at conducting her studies at Gombe: "Rusty the dog taught me that animals have personalities, minds and feelings of their very own."[9]

Along with two other girls who came to Bournemouth every summer, Jane and Judy formed the Alligator Club, which met in a small clearing near the house. They served tea, and sometimes met late at night for "feasts," which, because of the rationing of food that the war occasioned, often consisted of little more than a crust of bread or a biscuit. For Jane, these late night meetings were not so much about playing as about sneaking outside without being caught. The Alligator Club produced a small magazine that was filled with discussions of nature and anatomical drawings. On her own, Jane created a "conservatory" that contained flowers, shells, and even a human skeleton. She then charged admission. The monies she collected were donated to a local society that rescued old horses from being sold as meat and instead took them to safe havens where they could live out their remaining days.

Despite the disruptions of war, Jane's life at the Birches eventually fell into a comfortable routine. She grew close to her aunts and to her maternal grandmother. Raised a Congregationalist and having married a

Congregational minister and scholar, Danny Nutt tried hard to show her granddaughters the importance of Christian teachings. Hers was not a deeply intellectual faith; she emphasized instead the simple goodness of God and the importance of living a spiritual life. One of her favorite sayings, "As thy days, so shall thy strength be," would in time become a source of spiritual comfort for Jane.

REACHING OUT TO DESTINY

By the time she was eight years old, Jane had developed an unshakeable fascination with nature and animals. She read about them. She observed them. She drew pictures of them and wrote about them. In 1942 she received as a Christmas gift a copy of the children's classic *The Story of Dr. Doolittle*, a novel about a kindly doctor who develops the remarkable ability to talk to animals by learning their many languages. "I read it all the way through," she recalled, "then I read it again. That was when I first decided I must go to Africa some day."[10]

Jane also admired the Tarzan books, in which a young boy, raised by apes in the jungles of Africa, grows up to become the protector of the animals. She later wrote: "I was madly in love with the Lord of the Jungle, terribly jealous of his Jane. It was daydreaming about life in the forest with Tarzan that led to my determination to go to Africa, to live with animals and write books about them."[11] She also enjoyed reading other animal stories such as *The Wind in the Willows* and *Call of the Wild*. Another favorite, the Victorian tale *At the Back of the North Wind*, which told the story of a poor young boy living in a horse stable, impressed upon Goodall the extent of human suffering, a sad fact the war also brought home to her. Yet, these and other, more scientific books deepened Jane's knowledge of the natural world and whetted her appetite for further study. She had begun to dream about visiting Africa to investigate its wildlife. Making such a journey was, even now in the midst of her girlhood, becoming her heart's desire.

But it was a free book, purchased with coupons, that ignited her passion for learning. In a 2004 interview, Goodall recalled how she received the book:

> I had a nanny, I was about 6 at the time she stayed on when my sister was born. She saved up coupons. In those days you really got things free if you cut coupons off the packet of something, you didn't also have to send a check for 50 pounds as you do today. They say 'free' and it's not free at all. The prize was a hefty book, heavy,

dense, with photographs called *The Miracle of Life*. It was not for children. It went into the history of medicine and the discovery of anesthetics, and I can still see the pictures and I loved it. I read it, I drew it. I drew the insect mandibles. I really was a naturalist from the time I was born.[12]

"ALL CLEAR"

Increasingly, the droning of the German aircraft dropping their bombs intruded upon the peace of Bournemouth. While the Birches itself never was hit, Jane and her family spent countless hours huddled in a small air-raid shelter measuring six by five feet wide and four feet high, listening to the window glass shake and rattle. This small, steel-roofed structure was located in what had once been a servant's bedroom. Sometimes as many as six adults along with Jane and her sister crowded into the shelter, where they waited uncomfortably for hours until the "All Clear" siren indicated that the air raid was over and they could once more go about their business.

By the time Jane was seven, war had been raging in Europe for more than two years. She was by now familiar with news of the war and was coming to understand the immeasurable cruelty that humans could inflict upon each other. A narrow escape from German bombs during the summer of 1944 only increased her awareness of how uncertain and fleeting life could be. When news of the Nazi Holocaust against European Jews, accompanied by harrowing photographs of death camps, emerged, Jane was transfixed yet bewildered. The Second World War and the horrors it brought exerted a profound impact on Jane's mature view of the world. She often despaired of humanity and sought solace in nature among animals.

A CHANGING WORLD

In 1946, the year after the Second World War had ended, Mortimer Morris-Goodall divorced his wife. Jane was twelve years old and had become accustomed during the war to not seeing her father. With the divorce, however, her father and mother remained friends, and Jane kept in contact with her father throughout his life. Little else changed. At the time of her parents' divorce, Jane, her sister, and mother had been living at the Birches for almost seven years. Vanne's mother had since remarried, so for Vanne, her daughters, and her sisters, the Birches became truly their home and life continued on as before.

Jane started school, which she regarded as a necessary evil. Although she did enjoy learning about the subjects that interested her, such as the English language and literature, history, and biology, other subjects did not command her attention or induce her to work hard at her studies. The tedium of school did nothing to diminish her passion for reading and writing, which she continued to do on her own. Weekends and holidays were her escape. She spent time outdoors or curled up with a book. She had taken up writing poetry; many of her first efforts were about the joys of nature. Jane also continued to track local animals, keeping detailed notes of a hedgehog courting its mate, a weasel hunting mice, and a squirrel gathering nuts for the winter. Rusty still accompanied her on these outings.

She also began to talk more to Vanne about her dream to study wild animals in Africa. She also told her mother about the frequent taunts or scornful comments that she received from other students and teachers about her desire to study animals. Vanne told her never to take no for an answer. Once when Vanne was describing to her brother Eric her daughter's plans, Goodall heard her uncle say, "She doesn't have the stamina."[13] From that day on, Goodall, who suffered from migraines since she started school, never complained about her headaches again.

By this time, Goodall attended riding lessons on most Saturdays. Because she did not have the money to pay for the lessons, she cleaned saddles, bridles, and stalls. She was so dedicated and enthusiastic in her work that her teacher often gave her additional lessons for free and granted extra time to ride on her own. Jane soon began showing horses and participating in jumping competitions. She was an able rider and looked forward to her time at the stables.

When Jane got the opportunity to participate in a foxhunt, she was only too pleased to show off her horsemanship. She vowed not to let her teacher down. The day of the hunt found Jane in high spirits, too excited to think about the purpose of her ride. After riding for the better part of the afternoon, Jane finally spied the fox, exhausted by the long chase. Now she watched in horror as the hunting dogs set upon it, tearing its body apart. For Jane, the exhilaration came to a sudden and crashing end. Horrified and saddened by what she had just seen, she silently berated herself for participating in what she now regarded as a barbaric activity, just so she could display her horsemanship. That night, Jane lay awake in bed, haunted by the vision of the dogs and the fox. Even years later, she was angry with herself and could not forgive her lapse in judgment and character. Although Jane never gave up riding horses, she vowed never again to participate in a foxhunt or any other blood sport that resulted in the death of an animal.

GROWING UP

As she grew older, Jane increasingly preferred to spend time by herself. After school she sat alone in her grandmother's garden or did her homework in the summerhouse. She also spent a great deal of time in a special beech tree where she read or wrote. She was so attached to the tree she asked her grandmother to leave it to her in her will. Sitting in the branches 30 feet above the ground, Jane felt like a part of the tree and at one with nature. Sometimes when she laid her cheek against the bark, she thought she could feel the sap, the lifeblood of the tree, coursing through its limbs.

When Goodall turned 15 in 1949, her sporadic attendance at church became more regular. The reason for the change was the Reverend Trevor Davies, the new parson at the Richmond Hill Congregational Church. A Welshman, Davies was intelligent and a powerful speaker. Young Jane experienced her first crush, and over the next year she eagerly accompanied her grandmother to Sunday services.

But seeing Davies once a week was not enough. Soon Jane took to making excuses to take walks at night to stroll past his home. If she was lucky, she might catch a glimpse of him as he worked on his sermon. In an attempt to impress him, Jane started to read her grandfather's old theological books. Periodically, she went to the parsonage, rang the doorbell, and asked Davies for his opinion of a theological dispute about which she had read or to borrow one of the books from his personal library for additional study. Once he lent Jane a book on the philosophy of sensationalism in which the author argued that nothing existed outside the mind. For Jane, this was too much. She wrote Davies a humorous poem about the idea, which closed with the lines:

> And therefore I will cease to write
> Since I cannot be here,
> And none can ever read these lines
> For nobody is there![14]

To her dismay, Davies never mentioned her effort. Still, her infatuation grew. When the Reverend Davies shook her hand, she refused to wash it for days. On another occasion when Davies suggested to his parishioners that they go "the second mile"[15] in all they did, Jane took his message to a comic extreme. She began to fetch two buckets of coal and to brew two pots of tea. She even took two baths and told people "good night" twice. Her effort to live by the literal meaning of Davies's words nearly exasperated her family.

As many adolescents are wont to do, Goodall fantasized about living a saintly life and dying a martyr for her faith. She imagined traveling to the Soviet Union, which had banned all religious activity and imposed an official atheism. There she would join with the small groups of Christians and Jews who continued to meet despite the severe punishments the government inflicted for engaging in religious worship. Inflamed by the stories of the early Christians who met secretly in Rome after the government had outlawed their movement, Jane now saw herself as a missionary, bringing the Word of God to communist Russia. If caught, she would give up her life for her faith.

These religious fantasies of self-sacrifice and martyrdom were surely the heroic daydreams of an unusually thoughtful and sensitive teenager. But they cannot be so easily dismissed, for they reveal important facets of Jane's character. While still a young girl, she had developed a selfless devotion to the weak and the powerless. She showed a determination to battle injustice, untruth, and despotism. Perhaps most important, she struggled to change a world that she thought was increasingly dominated by barbarism, suffering, and evil.

In addition to reading the Bible everyday, Jane continued to indulge her love of poetry. She haunted the used bookshops, always on the lookout for affordable copies of her favorite poets: William Shakespeare, John Milton, Robert Browning, and John Keats. She was also fond of the English poets of the First World War, especially Rupert Brooke and Wilfred Owen. When not planning her martyrdom or, alternately, her trip to Africa, Jane dreamed of being a great poet, perhaps even one day becoming the Poet Laureate of England. Her poems were often humorous, but others incorporated her love of nature or explored her deepening spirituality. In "The Duck," for example, she wrote:

> The lovely dunes; the setting sun;
> The duck—and I;
> One Spirit moving timelessly
> Beneath the sky.[16]

By the time she entered her middle teenage years, Jane experienced a change in her spiritual life. Her growing sense of being one with nature, a theme she explored in "The Duck," gave rise to thoughts about the place and role of humanity in the world and to questions about whether human beings had the right to dominate nature just because they had the power to do so.

In the meantime, Jane had to attend to more practical matters that intruded upon her reflections. After graduating from high school in 1951, Jane, like many students, faced the problem about what to do. Her family did not have enough money to send her to a university and, although she was intelligent, her grades in the subjects for which she cared little now prevented her from winning a scholarship. As she wrestled with her future, there arrived the unexpected invitation from one of her aunts, her father's sister, to visit her and her husband in Cologne, Germany. Vanne agreed to accompany her daughter and together they set off.

For Jane, the trip was exciting yet dismal. The area she and Vanne visited was cold, bleak, and dreary. She did enjoy going with a young girl, Helga, to visit neighboring farms where she got to eat thick homemade bread and wear clogs. She enjoyed long walks where she stopped to watch hares making their way through the fields. Part of the purpose of the trip was for Jane to learn German. But the people around her were so anxious to learn English that she had little chance to practice their native tongue. It was just as well, for Jane realized that she was not particularly adept at learning languages.

One of the high points of her stay was visiting the great city of Cologne. Still scarred by the ravages of war, Cologne had already recovered some of its former charm. There was much to see and do. Jane wrote that gazing upon the spire of the great Cologne Cathedral, rising undamaged from the bombed ruins that surrounded it, brought home to her the ultimate power of good over evil. Vanne, being more practical, also hoped that Jane would master the German language in the hopes that it would help her daughter find work. Unfortunately, Jane showed no aptitude for it. Instead, Jane continued to plunge herself into seeing the sights. Yet, even amid the excitement of visiting a new country and seeing relatives, Jane could not escape her growing worries. Soon she and her mother would return to England, where she would have to confront anew the problem of what to do with the rest of her life.

NOTES

1. Jane Goodall with Phillip Berman, *Reason For Hope* (New York: Warner Books, 1998), p. 7.

2. Jennifer Lindsey and the Jane Goodall Institute, *Jane Goodall: 40 Years at Gombe* (New York: Stewart, Tabor & Chang, 1999), p. 19.

3. Jennifer Lindsey and the Jane Goodall Institute, *Jane Goodall: 40 Years at Gombe* (New York: Stewart, Tabor & Chang, 1999), pp. 19–20.

4. Jane Goodall with Phillip Berman, *Reason For Hope* (New York: Warner Books, 1998), p. 5.

5. Ron Arias, "Jane Goodall," *People Weekly*, May 14, 1990, p. 94.

6. "Jane Goodall," *Current Biography Yearbook, 1991* (New York: H. W. Wilson Company 1992), p. 249.

7. Jennifer Lindsey and the Jane Goodall Institute, *Jane Goodall: 40 Years at Gombe* (New York: Stewart, Tabori & Chang, 1999), p. 18.

8. Jennifer Lindsey and the Jane Goodall Institute, *Jane Goodall: 40 Years at Gombe* (New York: Stewart, Tabori & Chang, 1999), p. 18.

9. Steve Dale, "An Interview with Jane Goodall: What I Learned from Dogs," Studio One Networks, http://www.thedogdaily.com/you_dog/moments/archive/goodall_interview/(accessed November 15, 2004).

10. Allison Lassieur, "When I Was a Kid: Childhood Experiences of Famous People," *National Geographic World* (September 1999): 11.

11. Jane Goodall with Phillip Berman, *Reason For Hope* (New York: Warner Books, 1998), pp. 20–21.

12. Luaine Lee, "'Return to Gombe' with Jane Goodall on Animal Planet," *Knight Ridder/Tribune News Service*, Feb. 23, 2004, p. K4623.

13. Sy Montgomery, *Walking with the Great Apes: Jane Goodall, Dian Fossey, Biruté Galdikas* (Boston: Houghton Mifflin, 1991), p. 29.

14. Jane Goodall with Phillip Berman, *Reason For Hope* (New York: Warner Books, 1998), p. 23.

15. Jane Goodall with Phillip Berman, *Reason For Hope* (New York: Warner Books, 1998), p. 23.

16. Jane Goodall with Phillip Berman, *Reason For Hope* (New York: Warner Books, 1998), pp. 29–30.

Chapter 2

AFRICA CALLS

In the winter of 1952, 18-year-old Jane Goodall, fresh from her excursion to Germany, came home to the Birches where she faced an uncertain future. Unable to attend a university, and with no job prospects in sight, Jane was at a loss about what lay ahead. Recalling this uncertain period in her life, she wrote: "What would I do next? I only wanted to watch and write about animals. How could I get started? How could I make a living doing that?"[1]

Goodall knew that the time had come to put aside her dreams of traveling, at least for the moment. For the eldest daughter of a middle-class English family, there were more pressing matters at hand. She needed to find work to help support the all-female household at the Birches. Again, Vanne devised a solution: if Jane took classes to polish her typewriting, shorthand, and bookkeeping skills, she could find a job as a secretary. As Jane recalled, "Mum said secretaries could get jobs anywhere in the world, and I still felt my destiny lay in Africa."[2]

"LEARNING, LEARNING, LEARNING"

In May 1953, Goodall enrolled at the Queen's Secretarial College in South Kensington, just outside of London. She lived in the London home of Mrs. Hilliet, the mother of one of Vanne's friends. Goodall dutifully attended classes during the week, but she was bored. Writing to a friend, Goodall complained: "I'm very nearly dead. This shorthand is terrible hard work, and also rather monotonous as it only requires learning, learning, learning. The typing is not too bad, but that again, is a little bit automatic."[3]

Weekends were better. As soon as she could, Goodall escaped to the Chantry, a prosperous apple farm in Kent that was the home of another family friend. At the Chantry, Jane rode horses, took long walks, and relaxed.

Fortunately for Goodall, by April 1954 she had completed her training and returned to the Birches. In no time, she was back in her former routine. She rode horses whenever she could, and renewed her crush on the Congregationalist minister, Trevor Davies, though now her writing about him tended to be more ironic than romantic. All the while, though, Jane was thinking about how to realize her desire to journey to Africa and write about animals. That summer, Goodall confided in a letter to a friend that she had not given up her dream of becoming a journalist, but now thought that to "write anything worth anyone reading, I must have lived a few more years and acquired a little experience of life, as they say."[4]

In the meantime, Goodall earned money by helping her aunt Olly, a physiotherapist, at a local clinic. Many of Olly's patients were young children who had been paralyzed by polio, or who suffered from such crippling ailments as muscular dystrophy and cerebral palsy. Goodall transcribed Olly's comments on each case and then typed and filed the reports. Although she dreamed of going to Africa and pursuing a career in journalism, Goodall's time at the clinic was not misspent. Working day after day with the ill and infirm, especially children, Jane's empathy grew for those who were weak and disabled. In fact, her first boyfriend was a young man injured in a terrible car crash and in a cast from his waist to his ankles. When working at her aunt's clinic, Goodall visited her uncle, who was a surgeon and who let her watch him as he operated. As a result of these experiences, Goodall developed a new sense of compassion and a deep gratitude for her good health. She also came to marvel at the indomitable spirit of those who faced serious, debilitating, and often painful injuries or illnesses.

OXFORD

In August 1954, Goodall took advantage of an opportunity to move to Oxford to work as a typist for the Oxford University Registry. Like her other secretarial jobs, this one did not provide much stimulation. But it enabled Jane to earn a decent income and to work in Clarendon House, one of the more interesting buildings at Oxford University. Built during the early eighteenth century, Clarendon House boasted grand double entrance doors, large Doric columns, and magnificent statues that overlooked the grounds. Jane liked going to work in such an elegant setting.

She enjoyed bringing her pet hamster, Hamlette, with her everyday. Yet, Clarendon House and Hamlette were not enough to alleviate the drudgery of her work. She wrote to a friend that "I have been miserable these last few weeks because of the boredom of this foul job."[5]

Still, she did enjoy living in Oxford. Her residence was a boarding house located at 225 Woodstock Road. She became close friends with her roommates, all of whom were single women near her own age. In addition, she met other congenial young men and women at work. Things went well enough for Jane in Oxford. She had a decent job and pleasant companions. But Jane soon grew frustrated and restless. Nothing it seemed could dispel her feelings of uncertainty about the future.

TO LONDON

By late July 1955, Goodall had moved once more, this time to London where she took a job at Schofield Productions, a company that made documentary films. Goodall's job at Schofield was more interesting than her former position: she selected the music for the films. The job offered a welcome break from her dull secretarial duties. At Schofield, Goodall learned a great deal about making films, knowledge that later proved invaluable to her. In addition, she made the most of living in London. She took classes, attended lectures, went to concerts, and enjoyed an active social life. Goodall enthusiastically wrote about the many gentlemen who called on her, sometimes referring to them by their full names (David, Keith, Horst, and Hans). She referred to those who became closer to her simply by using a letter, such as "B." While in London, Jane also saw more of her father; her letters home speak of their going to the theater and to restaurants.

For the first time in a long time, Goodall seemed content with, and even enthusiastic about, her life. But in May 1956, Goodall's fortunes took an unexpected turn, and she realized her dream of working with, and writing about, animals might come true.

KENYA CALLING

By the morning post on Wednesday, December 18, 1956, Jane received an interesting letter. Mailed with stamps that depicted elephants and giraffes, the letter had come from an old school friend, Marie Claude Mange, who had moved to Kenya with her parents and had recently bought a farm. Mange wrote to ask whether Jane would be interested in coming to see her in her new home. Goodall was dumbstruck. She could not believe her good fortune. For so long, she had wanted to go to Africa, and now, quite unexpectedly, an opportunity to do so had come. Jane did not hesitate. She was going.

The problem was that traveling to Africa cost money, and Jane could not afford to make the trip. She also knew that she had to raise the price of a round-trip ticket, for Vanne would never let her go if she thought Jane had no plans to return to England. In any event, neither English nor Kenyan authorities permitted visitors to enter Kenya with a one-way ticket, unless someone assumed legal responsibility for their welfare. Not wishing burden Marie and her family, Goodall made up her mind to raise the money she needed to purchase a round-trip ticket.

Jane gave notice that she intended to quit her job at Schofield shortly after receiving Marie's letter. She returned to the Birches where she could live rent free. She took a job as a waitress in the Hawthorns, a local hotel. In a letter written during late summer 1956, Jane explained: "I am working myself absolutely to the bone. It really is dreadful during the peak of the season. We only get one day off a fortnight [two weeks], two afternoon teas and one late night per week."[6] Yet, despite her unhappiness with the job, Goodall was saving a lot of money. With each paycheck and every tip, she moved a little closer to Kenya. She hid her money under the carpet. After five months, Jane and her family shut the curtains one evening and pulled up the carpet to see what she had saved. Much to her delight, Jane found that she had more than enough money to pay for a round-trip passage to Kenya. She was nearly on her way.

Before she left, however, Jane had to disentangle herself from her London beaus. In letters to her friends, she describes having to reject at least two proposals of marriage. It was now that Jane began the slow retreat from her familiar life and instead began to concentrate on what became her destiny.

SAILING TOWARD DESTINY

"It is now 4 P.M. on Thursday," wrote Goodall on March 15, 1957, "and I *still* find it difficult to believe that I am on my way to *Africa*. That is the thing—AFRICA."[7] Now that Jane's dream was coming true at last, it seemed more than ever like a dream to her. That "the adventure, the voyage to Tarzan's Africa, to the land of lions, leopards, elephants, giraffes, and monkeys had actually begun,"[8] that she was really on her way to Africa, seemed incredible. All Jane's efforts, however, might have come to nothing, for at the time she planned to leave, Great Britain and Egypt were at war. As a result, the Suez Canal, through which her vessel would have to pass, closed the week before Jane was to sail. There was a possibility that the voyage would be canceled. To her great relief, Jane learned that the company decided to go forward with the trip, though she had pay more for her ticket and spend an extra week at sea. For Jane, the extra money and the inconvenience were worth it.

Her send-off had been both a happy and sad occasion. Vanne and her Uncle Eric came to bid Goodall bon voyage. The *Kenya Castle*, on which she had booked passage, was a large passenger liner of the famed Castle Line. Goodall liked the ship because it was one of the few that did not separate traveler's quarters into steerage and first class. Although she had to share a small stateroom with five other young women, these arrangements could not dampen Jane's excitement. In three short weeks, she reflected as the ship set sail, she would arrive in Africa. At the age of 23, Goodall sensed that this journey involved more than a reunion with a former classmate. It was, rather, the turning point of her life.

Goodall enjoyed the voyage. When the *Kenya Castle* made for open sea, she stood in the bow of the great ship and looked out over the water that stretched in endless waves toward the horizon. She even liked the stormy weather, and ventured out on deck to feel the sea spray against her face. She made friends with her cabinmates and indulged in a few shipboard flirtations. Years later, Goodall admitted that while the human faces from that voyage had grown somewhat hazy, she still remembered clearly her moods while watching the ocean, the sky, and the sea life. She also recognized something important was happening to her: "I think it was then, sailing along just out of sight of land, that I made an unconscious commitment to Africa. The days of my childhood, and of my adolescent preoccupation with philosophy and the meaning of life, of time, of eternity, had come to an end."[9] Goodall thought about all she had learned growing up and began to see how her upbringing, her schooling, her deep love for the natural world, and even the horrors of the war had readied her to enter the most significant stage of her life. She had no idea of what lay ahead, but felt a growing confidence that she could meet the new challenges without fear.

On the journey to Africa, the *Kenya Castle* visited four ports: the Canary Islands, Cape Town, Durban, and Beira. At each port of call, Goodall was taken by the exotic food, the heat, the markets, and the people. But her experience was also tempered by a cruel dose of reality. While in Cape Town, Goodall came face to face with apartheid, a practice in South Africa that imposed strict segregation between blacks and whites. Everywhere she looked she saw signs that read in Dutch "SLEGS BLANCS" ("WHITES ONLY"). The signs reminded her of the treatment the Jews had received in Germany and elsewhere in Europe during the 1930s and the dark years of the Second World War. She never forgot either experience.

During the early hours of April 2, 1957, the *Kenya Castle* chugged into the port at Mombasa, Kenya. For the next several hours, Goodall made her way through customs and by noon was on a train for Nairobi station, a daylong journey from Mombasa. She was as entranced by the landscape of East Africa as she had been by the sea. On the morning of April 3, which also happened to be Goodall's 23rd birthday, she stepped off the train in Nairobi where her friend, Clo Mange, Clo's father, Roland, and another friend named Tony were waiting to meet her. After gathering Goodall's belongings, they set out for the farm. Driving north, they soon entered the White Highlands. Paved highways gave way to dirt roads. The trip was long and uncomfortable, but at last the travelers arrived at the small trading center of Naivasha and the greystones where the Mange family farm was located. Goodall could hardly believe her eyes when she spied a giraffe standing beside the road.

AN AFRICAN LIFE

For the next few weeks, Goodall stayed with the Manges on their farm. She tried to make the most of every moment; she took in the crisp mountain air and the beautiful clear streams. She was enthralled by the many different kinds of birds and was excited to see the footprints of a giant leopard. For Goodall, Africa was almost a magical experience. As a young girl, she had dreamed of the moment when she set foot in Africa. Now, more than a decade later, she was standing on African soil.

Unfortunately, Goodall's efforts to view African wildlife were hampered by political chaos and bloodshed. Kenya was under British rule. But by the late 1940s, many native Kenyans grew increasingly dissatisfied with being subjects of the British Empire and began to call for independence. Violence exploded throughout the country. On October 7, 1952, Chief Kungu Waruhiu, a strong supporter of the British, arrived at the Seventh Day Adventist mission seven miles outside of Nairobi. Moments later, bullets riddled his car and killed him. The gunmen were Mau Mau rebels, members of a secret society who had vowed to free Kenya from British rule and drive the white man out. Already suspect of committing arson and slaughtering cattle belonging to British ranchers, the Mau Mau's dramatic assassination of Chief Waruhiu stunned the British colonial government, which, in response, declared a state of emergency in Kenya that lasted nearly eight years.

The home government in England dispatched troops to help quell the violence and maintain order. Colonial authorities detained more than 100,000 Kenyans in detention camps and ordered troops to hunt down and capture members of the Mau Mau. Despite these efforts and precautions, the violence continued and, in fact, accelerated. The Mau Mau engaged in horrific massacres, usually among those living in the more remote highland regions of central Kenya. Yet, only 32 Europeans died as the result of these attacks. The principal targets of the Mau Mau's rage and vengeance were the more than 2,000 Kikuyu, a native Kenyan people who were loyal to, and who cooperated with, British authorities. It took another six years of bloodletting before the British government granted Kenyan independence. The brutality and horror of these events affected Goodall deeply, as much, certainly, as the stories of the Holocaust. Goodall never ceased to be amazed at the pain human beings willfully inflicted on one another.

A SHAMEFUL EXPERIENCE

During her time with the Manges, Jane decided to do something she swore she would never do again: go hunting. She later asked herself why she had agreed to go along with an activity that she clearly found shameful and despicable. Yet, she had to admit, at least to herself, that her attraction to a local young man fueled her participation in the hunt. Trying vainly to impress him, Jane asked to ride his horse, which had a reputation for being difficult to handle. The horse had already thrown a number of riders more skilled than Jane, and was one of those horses that choose their own master, consenting only to be ridden by his owner. Jane persisted, and the young man finally but reluctantly agreed.

Jane mounted the horse, which at more than six feet tall, was the biggest horse she had ever ridden. She set off with a group of other riders for what she thought was nothing more than a ride through bush country. Not until it was too late did Goodall realize that she was on a hunt for jackal. To her immense relief, the hunters had bad luck that day and shot nothing. Nonetheless, Jane was ashamed, even unwittingly, to have again taken part in an activity that she had come to condemn and loathe. Angry at her vanity, Goodall never participated in another hunt.

A STEP CLOSER

In the three weeks since Jane had arrived in Kenya, she busied herself exploring the countryside and observing the animals. Jane, though, often had men on her mind. As a young woman, she was something of a flirt and found no shortage of men eager to call on her. She even received, and evidently entertained, one serious proposal of marriage. Jane also moved in a privileged social circle. The primary focus of the group was on riding, breeding, and racing horses, and Jane quickly established herself as a capable rider. But as for her romantic adventures, Jane never thought of marrying a rich man and becoming part of the horsing set in Kenya. She enjoyed herself, but was consumed with other ambitions.

In the fall of 1957, after Goodall arrived in Africa, Vanne wrote her daughter a letter: "Sometimes now I feel you are utterly lost. That great gorgeous primitive continent has swallowed you whole—you are engulfed in huge clouds of heat—stolen by a thousand alien voices—utterly remote from this tiny grey island where cold winds take the warmth from the sun."[10] By now Goodall knew she wanted to stay in Africa and, in some capacity, to work with and write about animals. Not wishing to take further advantage of her hosts' hospitality, Jane wanted to find a place of her own and begin looking for a job. So it came about that she left for Nairobi.

The capital of Kenya, Nairobi is situated in the southern highlands. Not surprisingly, it is the economic and cultural center of the country, in addition to being the hub of its political life. The manufacture of textiles, clothing, and transportation equipment, along with the processing of food, dominated the economy of the city, which also depended on an extensive tourist trade.

Historically, Nairobi was part of an area once dominated by the Masai, a nomadic people. British colonists actually established the modern city in the late nineteenth century as a railroad stop on the Mombasa-Uganda line. Between 1899 and 1905, Nairobi served as the British provincial capital for the region. In 1905, the city became the capital of the entire British East Africa Protectorate. By the time Goodall arrived in the late 1950s, the city was also known as Kenya Colony. A few years later, in 1963, Nairobi was still the capital, but this time of an independent Kenya.

Thanks to her Uncle Eric, Jane had managed to secure a job as a secretary for the Kenya Branch of a major British company. She found the job to be extremely boring, but it provided a steady paycheck, allowing her to stay in Kenya. Her living quarters were modest. To keep expenses to a minimum, she stayed in a local hostel that was cheap but comfortable. Yet, she dreamed of finding a job that would fulfill her desire to work with animals.

Within two months of taking up residence in Nairobi, Jane once more enjoyed a busy social life that consisted of riding, attending social events, and going to dinner parties. At one of these numerous dinner parties, her life took another of its unexpected, but fortunate, turns. After explaining her desire to work with animals, a friend suggested: "If you are interested in animals, you should meet Louis Leakey."[11]

NOTES

1. Jane Goodall, *My Life with Chimpanzees* (New York: Simon and Schuster, 1996), p. 35.
2. Ron Arias, "Jane Goodall," *People Magazine*, May 14, 1990, p. 94.
3. Jane Goodall, *Africa in My Blood* (New York: Houghton Mifflin, 2000), p. 39.
4. Jane Goodall, *Africa in My Blood*, ed. Dale Peterson (New York: Houghton Mifflin, 2000), p. 39.
5. Jane Goodall, *Africa in My Blood*, ed. Dale Peterson (New York: Houghton Mifflin, 2000), p. 40.
6. Jane Goodall, *Africa in My Blood*, ed. Dale Peterson (New York: Houghton Mifflin, 2000), p. 72.
7. Jane Goodall, *Africa in My Blood*, ed. Dale Peterson (New York: Houghton Mifflin, 2000), p. 76.
8. Jane Goodall with Phillip Berman, *Reason for Hope*(New York: Warner Books, 1999), p. 38.
9. Jane Goodall with Phillip Berman, *Reason for Hope* (New York: Warner Books, 1999), p. 40.
10. Jane Goodall, *Africa in My Blood*, ed. Dale Peterson (New York: Houghton Mifflin, 2000), p. 7.
11. Jane Goodall with Phillip Berman, *Reason for Hope* (New York: Warner Books, 1999), p. 44.

Chapter 3

A MOMENTOUS MEETING

Louis Leakey was among the most celebrated but also the most controversial paleoanthropologists in the world. A pioneer in the field of paleoanthropology, the study of early human beings and their antecedents, Leakey based his work on the excavation of fossilized remains and cultural artifacts, such as stone tools. Leakey's research helped to change the view of how humans originated and evolved. The main focus of Leakey's work was the search for clues about how and when humans and apes split off from their common ancestor.

THE MAKING OF A GENIUS

Leakey had spent most of his life in Kenya. Born on August 7, 1903 in Kabete, Kenya, Leakey's parents were Anglican missionaries. As a child, Leakey learned the Kikuyu language and customs. His knowledge later enabled him to compile a primer of Kikuyu grammar. His interest in natural science and anthropology also began during his youth. As a boy, he took up ornithology, the observation and study of birds. During his bird-watching expeditions, he often came across stone tools washed out of the soil by heavy rains. Youthful excitement notwithstanding, Leakey suspected these tools were of prehistoric origin.

During the early twentieth century, archaeologists believed that the presence of such stone tools as Leakey had found was the primary means of proving the existence of humans, because tool-making was an exclusively human activity. However, scientists did not consider East Africa a likely site for finding evidence of early humans. Discovery of the so-called Java

Man in 1894 led scientists to assume that Asia was the home of the first *Homo sapiens*. In time, Leakey's studies came to question that assumption.

FIRST DISCOVERIES

To complete his formal education, Leakey went to England where he studied archaeology and anthropology at Cambridge University beginning in 1922. A rugby injury caused him to take a leave of absence from school, but the injury brought an unanticipated advantage. Leakey spent his time away from Cambridge traveling with a British Museum archaeological expedition to Tanganyika (now Tanzania). The trip was a major turning point for Leakey. As a result of the expedition, he became convinced that there was much more to learn in East Africa about the origins of human beings. But Leakey might as well have been alone in the African desert. Not one of his professors agreed with his theory, trying instead to convince him that Asia, not Africa, was the birthplace of the human race. Leakey stubbornly clung to his idea, and after graduating in 1926 with degrees in anthropology and archaeology, he spent the late 1920s and early 1930s traveling to East Africa, and to Kenya in particular, to continue field research for his doctoral dissertation.

Leakey's hunch paid off. In 1932, near Lake Victoria, Kenya, Leakey found the remains of *Homo sapiens* (modern man), discovering the so-called Kanjera skulls that were more than 100,000 years old. He subsequently found a Kanam jaw, which was more than 500,000 years old.

When Leakey returned to England, many archaeologists praised his findings. Yet, others questioned them and soon a major scholarly controversy was brewing. Leakey responded to his critics by inviting the noted English geologist Percy Boswell to visit the sites in 1934. This decision proved unfortunate. Boswell found Leakey's documentation inadequate, and suggested that Leakey could not reliably validate either find. Boswell's report seriously damaged Leakey's scientific reputation. But even with this setback, Leakey refused to give up hope that he would discover solid evidence of the existence of *Homo sapiens* from the Lower Pleistocene period, which dated from between approximately 1.8 million years to 10,000 years ago.

A NEW PARTNER

During this period, too, Leakey's personal life fell into turmoil. In 1933, Leakey met Mary Douglas Nicol and fell in love with her. At this time, Leakey had been married for five years. He and his first wife already had one child and had another on the way. The situation was tense and delicate.

Mary Nicol had spent much of her childhood traveling throughout Europe. She had visited several prehistoric sites, such as the caves at Pech Merl in Dordogne, France. This life of traveling influenced her decision to become a geologist and an archaeologist, which was a bold move for a woman at the time. Nicol was also blessed with artistic ability. By the time she was 17 years old, Nicol had found work as an illustrator at the Hembury Dig in Devon, England. For the next two years she worked at the dig illustrating the artifacts that the team unearthed. Nicol had an intense interest in the Stone Age; the drawings from her time at Hembury, especially her skillful renderings of Stone Age tools, displayed the great talent she possessed.

Despite being a married man, Leakey took Mary along on his next expedition to Africa in 1934. After the Boswell debacle, the couple returned to England the following year, and soon Mary was living with Leakey. A year later, in 1936, Leakey's wife Frida filed for divorce. Louis and Mary married later that year. Leakey may have been happy in his personal life, but the affair and the divorce had wrecked him professionally. The scandal destroyed any chance that Leakey may have had of pursuing an academic career. The discoveries at Kanam and Kanjera, which Percy Boswell had so brutally shot down, also helped to tarnish his reputation, perhaps beyond repair. So, without a steady job, Leakey earned a small income from speaking and writing. In 1937, he and Mary decided to return to Africa.

Although begun under difficult and, by the standards of the day, embarrassing circumstances, Leakey's marriage to Mary Nicol became a true partnership. She soon distinguished herself as one of the most important members of Leakey's archaeological team and is credited with making several important discoveries of her own.

UNLOCKING THE HUMAN PAST

By the end of the 1930s, Leakey had altered the focus of his research. Instead of concentrating his efforts on the discovery of stone tools as evidence of human existence, he concentrated more fully on human and prehuman fossils. His studies led him and his team to Rusinga Island, located at the mouth of the Kavirondo Gulf in Kenya. During the late 1930s and early 1940s, Leakey's team uncovered on Rusinga a large number of fossils, particularly the remains of Miocene apes. The discovery was crucial. Named *Proconsul africanus*, the remains showed a jaw that was more human than ape. For Leakey, the find provided additional evidence that the *Proconsul* represented a stage in the evolution from ape to man.

In 1941, Leakey was made an honorary curator of the Coryndon Museum. Four years later, in 1945, he accepted a poorly paid position as curator of the museum so he could continue his work in Kenya. In 1947, Leakey organized the first Pan-African Congress of Prehistory, which introduced many scientists to the work that he and his team had accomplished since the Kanam/Kanjera debacle of 1934. The event turned out to be a professional and personal triumph, which aided immeasurably in restoring Leakey's reputation in the scientific community.

In 1948, Mary Leakey unearthed a nearly complete *Proconsul* skull, which dated back at least 20 million years. The find was monumental. It was the first fossilized ape skull ever discovered. Today, most archaeologists consider the *Proconsul* too specialized to have been a direct ancestor of human beings; however, the *Proconsul* remains important in the quest to explain the origins of life on earth.

Four years later, in 1952, Leakey began his excavations at Olduvai Gorge, Kenya. When the Mau Mau uprising forced him to suspend his work, he turned his energies instead to writing a book called *Mau Mau and the Kikuyu*. In this book, Leakey attempted to explain the Mau Mau rebellion from the perspective of a European raised in Kenya who happened to possess an extensive knowledge of Kikuyu language and culture. Leakey also published a second work, *Defeating Mau Mau*, in 1954.

Louis Leakey was once asked in a radio interview what kept him going all these years. His reply was simple: "I want to know: Who am I?"[1] That simple question eventually led Leakey seriously to consider the relationship between primates and humans.

A FATEFUL MEETING

Not long after her dinner party conversation, Jane Goodall took her friend's advice and set out to meet Louis Leakey. In a 2004 interview, she recalled their initial meeting. "I'd heard about [Leakey] and called up and said, 'I want to come and see you.'"[2] Leakey's initial response to Goodall: "I'm Dr. Leakey. What do you want?"[3] Still determined to meet Leakey, she made an appointment to visit him in his office at the Coryndon National Museum (now called the National Museum) in Nairobi. At ten o'clock on the morning of May 24, 1957, she stepped into Leakey's large and untidy office, which she later described as strewn with piles of paper, fossils of bones and teeth, primitive stone tools, and a host of other artifacts.

The two immediately began a lively conversation as they walked about the museum, with Leakey from time to time pausing to point out some of the more interesting exhibits. Goodall was anxious about meeting

Leakey, but to her relief she managed to answer most of his questions. She also thought that Leakey was impressed with the knowledge of Africa and animals that she had acquired on her own: "I hadn't been to university. I said I've saved up, [and] am staying with a friend. And I had a temporary job and I really wanted to work with animals and could he help me? He gave me a job instantly as his assistant."[4]

As it turned out, Leakey's secretary, Rosalie Osborn, had planned shortly to leave her position. Leakey hired Goodall as her replacement. Unfortunately, Goodall soon learned that Osborn had quit because of one of Leakey more unattractive qualities: his relentless pursuit of women, despite his own marital status, or theirs. Drawn to young, attractive women, Leakey had a reputation for developing close and informal relationships with the female protégés who came to work with him. Despite Mary's jealousy and anger, Leakey continued his behavior. His relationship with Osborn, however, became serious and, as one of his sons later remarked, Osborn almost became "the third Mrs. Leakey."[5] By 1955, quarrels between the Leakeys had become more frequent and altercations were commonplace.

When Jane arrived at the museum, Mary was convinced that Leakey had once again taken up with another of his female assistants. She was right. Leakey soon expressed a romantic interest in Jane—an interest that Jane not only did not reciprocate but that also left her frightened and dismayed. She did not want to lose her job and endanger the possibility of working with animals. Although Leakey persisted in making unwanted advances, he finally realized that nothing he could do was likely to convince Jane to pursue the affair. To his credit, instead of firing Jane, he agreed to keep her on, and, if she were willing to continue their association, to serve as her mentor.

For the next year, Goodall worked in Leakey's office at the museum without incident. She made the most of her time there, continuing to learn about the animals, peoples, and culture of East Africa. For his part, Leakey was charmed by Goodall and captivated by her enthusiasm.

During her time at the museum, Goodall began to accumulate an exotic menagerie of animals, including two "bush babies," agile, long-tailed nocturnal African lemurs with dense, woolly fur and large eyes and ears, a hedgehog, several dogs, cats, and fish, a mongoose, a monkey, and an assortment of snakes, spiders, and rats. Later, however, Goodall campaigned against people owning exotic pets, stating that wild animals belonged in the wild. To domesticate these animals did them a disservice. But at that period in her life, Jane's pets symbolized her attempt to live among the wild beasts. The problem was that she had not quite figured out how to do so.

MEETING THE SERENGETI

During the late 1950s, the Leakeys continued their work at Olduvai and, in 1957, invited Goodall along with Gillian Trace, another English girl who worked at the museum, to accompany them on their next dig. Goodall was excited at the prospect of participating in her very first archaeological dig.

Every year for three months, the Leakeys dug at Olduvai, searching for human remains. Over the years, Louis and Mary had increased their knowledge about the great prehistoric creatures that had once roamed the area. But so far, little had come of their expeditions; they had coaxed only simple stone tools from the earth. Still, year after year the Leakeys returned, convinced that the earth would eventually yield the bones of ancient man.

The region surrounding Olduvai was known as the Serengeti, a remote area that the Masai people called "the place where land moves on forever."[6] The Masai was the only group to inhabit the area. A nomadic people, they moved from place to place on the seemingly endless grassy plains. Outsiders had come to the Serengeti only in 1913, when Stewart Edward White, an American hunter, journeyed to the region. Pushing south from Nairobi, White recorded: "We walked for miles over burnt out country.... Then I saw the green trees of the river, walked two miles more and found myself in paradise." White became the first white man to set foot on Serengeti. In the years since White's excursion, the Serengeti came to represent paradise to any who visited it.

But in 1957, more than 40 years after White's expedition, Olduvai remained isolated from the rest of Africa and the rest of the world. The Masai still lived in the area. As Goodall soon found out, traveling to Olduvai thus posed several problems. Because the area was so remote, there were few roads and none that lead directly to the dig. When Leakey's team left the village of Ngorongoro on the Seronera Trail, therefore, Jane and Gillian had to sit atop the Leakeys' Range Rover to look for signs of tire tracks from the previous year so the driver knew what route to follow.

To complicate matters, the team had to contend with Mary's increasingly erratic behavior on a journey that was already long and tiring. Mary drank more heavily than she had previously done. Often intoxicated, Mary required special looking after, a responsibility that often fell to Jane. Once the team arrived at Olduvai Gorge, the work of setting up camp provided a welcome distraction. Under the cover of the shady acacia trees, Jane helped to erect the three small tents that served as the group's home for the next three months. Soon thereafter a truck arrived carrying additional equipment and bringing the African staff that worked with the

Leakeys on all their expeditions. Once emptied, the truck served as sleeping quarters and a small office for the couple.

At night, after dinner, Jane lingered around the campfire listening to the sounds of the Serengeti. She heard the roar of a lion and the high-pitched giggle of a hyena. For Goodall, her dream was finally coming true. Miles from civilization, Africa enveloped her. It was one of the happiest moments in her life.

FIELD EXPERIENCE

At Olduvai, Goodall's primary job was looking for fossils. In many respects, the work was as monotonous as it was exhausting, made even more so by the necessity of working in the blistering sun. Still, Jane was fascinated. Feeling mysteriously transported to a primeval world that was older than human memory, she later wrote:

> I will always remember the first time I held in my hand the bone of a creature that had walked the earth millions of years before. I had dug it up myself. A feeling of awe crept over me. I thought, "Once this creature stood here. It was alive, had flesh and hair. It had its own smell. It could feel hunger and thirst and pain. It could enjoy the morning sun."[7]

Before digging could begin, the Africans removed the topsoil using picks and shovels. Once near the fossil layer, Mary Leakey supervised the work of the diggers, stating that if an important fossil were damaged, she would assume the blame. When the fossil bed was located, Mary and Jane dug into the site using hunting knives to chip away the hard soil. After locating a fossil, they used more delicate tools, such as dental picks, gently to extract the item from the earth. Pausing for a cup of coffee at eleven o'clock and for a three-hour rest during the hottest part of the day, Jane and Mary worked tirelessly.

A POWERFUL MENTOR

After Leakey realized that Goodall had no romantic interest in him, the two soon settled into a relationship of teacher and student. In her autobiography, published in 1999, Goodall states that Leakey had a tremendous influence on her life, both as a naturalist and as a person. Among the endearing qualities for which Jane was so grateful, and from which she benefited so much, was Leakey's love of conversation. Leakey was a masterful storyteller and also loved to talk about his work, especially the stone tools he had found and studied over the years. He often demonstrated for Jane how the tools were constructed and used, and speculated about when they had been made. For Jane, though, animal fossils became more interesting than tools. During that first dig at Olduvai, she came to examine them more and more carefully. Her growing interest in animals paralleled Leakey's ongoing quest to discover the origins of creatures that could be classified as distinctly human. Where did the animal end and the human begin? That was the question.

HUMAN ANCESTORS?

For Leakey, the study of chimpanzees, gorillas, and orangutans held one of the keys to unlock the mystery of human origins. Chimpanzees were of particular interest. Found only in Africa and living in an area that stretched from the equatorial forests to the west coast and then ran eastward to Uganda and Tanzania, chimpanzees also were found in the rugged mountainous area along the shore of Lake Tanganyika, which was approximately six hundred miles southwest of Olduvai. These creatures were known as *Pan Troglodytes schweinfurthii* and were distinguished by their long hair.

Leakey believed the study of chimpanzees was of immense importance to understanding not only the appearance but also the behavior of early man. For Leakey to prove his theory, which at the time did not garner much support from other anthropologists, archaeologists, and geneticists, required extensive field studies. But at that point in his career, Leakey had no intention of abandoning his study of fossils and artifacts to carry out the work himself. To be sure, the task was daunting. No one had ever undertaken the kind of study Leakey envisioned. As a consequence, researchers had no guidelines to follow. In addition, the work would have to be conducted in the remote habitat of the chimpanzees and require observations to be made over a long period of time. Leakey could think of only one person who was up for the job, and he began to formulate a plan.

BACK TO NAIROBI

When the three-month expedition to Olduvai came to an end, Goodall, the Leakeys, and the rest of the team returned to Nairobi. Jane resumed her duties at the museum, but grew increasingly discontented and unhappy. She found that she liked fieldwork, and wanted more than ever to work with animals in some capacity.

Jane's personal life was more satisfying. Vanne came for a visit and, like her daughter, immediately fell in love with Africa. She soon made friends of her own and began taking trips to see as much of Kenya as she could. Jane was also dating a man who, ironically, escorted clients on hunts for big game. Despite his business, Jane enjoyed his company and saw admirable qualities in him. He was brave and had overcome personal adversity. He also excited Jane's compassion in the way her aunt's patients had done. When Jane met him, he was in a plaster cast from his feet to his waist, the result of a severe auto accident months before. During their entire time together, he was lame. He was also kind to animals, which raised him further in Jane's estimation. The romance ended after about a year. As the hunter recovered from his injuries and began to think about returning to his work, Jane decided that she simply could not accept what he did for a living.

After her breakup, Jane fell into a mild depression. Her present life and her future prospects seemed bleak. She recalled:

> I could have gone on at the museum. Or I could have learned a whole lot more about fossils and become a paleontologist. But both these careers had to do with dead animals. And I still wanted to work with living animals. My childhood dream was as strong as ever:—Somehow I must find a way to watch free, wild animals living their own, undisturbed lives.—I wanted to learn things that no one else knew, uncover secrets through patient observation. I wanted to come as close to talking to animals as I could.[8]

ANOTHER STEP FORWARD

Leakey continued to mention the need for an extensive field study of chimpanzees. Jane required no encouragement, for she wanted desperately to undertake a study of this nature. One day, she went into Leakey's office and "blurted out:"

> "Louis, I wish you wouldn't keep talking about it [the field study] because that's just what I want to do."
>
> "Jane," he replied..., "I've been waiting for you to tell me that. Why on earth did you think I talked about those chimpanzees to you?"[9]

Goodall was stunned. She had no academic training or credentials; there was no way she thought herself suitable to undertake such a project, however much she may have wanted to do so. But Leakey reassured her about her capabilities and told her that she was the perfect candidate for the job. Isolated from scholarly prejudices, she would approach the project with an open mind. Goodall had already demonstrated her ability to live in primitive conditions. She also possessed a deep love of animals and had a great deal of patience, qualities that would stand her in good stead during the long months of isolation. Leakey believed all these traits more than made up for her lack of academic credentials. Yet, he admitted that though he had been planning to propose this course for a long time, he had been cautious about approaching her. Leakey wanted to make sure during their conversations that Jane fully appreciated the requirements and the risks that such an undertaking entailed. He had hoped, in fact, that she would come to him, and now she had.

Jane was overjoyed and ready to leave that very day. But Leakey prevailed upon her that there was much work to be done before she was off for the mountains. Among other tasks, Leakey still needed to raise the money to fund the project and also needed permission from British colonial authorities as well as local authorities in Kenya and Tanzania before he could allow her to proceed. Making the necessary arrangements, he explained, would take time, perhaps as long as a few years. In the meantime, Leakey advised Goodall to learn more about chimpanzees. The best place to accomplish that, he told her, was not in Africa but in England.

NOTES

1. Sy Montgomery, *Walking with the Great Apes: Jane Goodall, Dian Fossey, Biruté Galdikas* (Boston: Houghton Mifflin, 1991), p. 74.

2. Luaine Lee, "'Return to Gombe' with Jane Goodall on Animal Planet," *Knight Ridder/Tribune News Service*, Feb. 23, 2004, p. K4623.

3. Sy Montgomery, *Walking with the Great Apes: Jane Goodall, Dian Fossey, Biruté Galdikas* (Boston: Houghton Mifflin, 1991), p. 79.

4. Luaine Lee, "'Return to Gombe' with Jane Goodall on Animal Planet," *Knight Ridder/Tribune News Service*, Feb. 23, 2004, p. K4623.

5. Virgina Morell, *Ancestral Passions: The Leakey Family and the Quest for Humankind's Beginnings* (New York: Simon & Schuster, 1995), p. 244.

6. "Discover the Serengeti," http://www.serengeti.org/ (accessed November 1, 2004).

7. "Jane Goodall," http://www.janegoodall.ca/jane/jane_bio_early.html (accessed October 10, 2004).

8. "Jane Goodall," http://www.janegoodall.ca/jane/jane_bio_early.html (accessed October 10, 2004).

9. Jane Goodall with Phillip Berman, *Reason for Hope* (New York: Warner Books, 1999), pp. 54–55.

Chapter 4

THE HIDDEN WORLD OF THE CHIMPANZEE

By the time Goodall left Kenya in 1958, Leakey was already hard at work to line up funding for the proposed chimpanzee field study. The study would take place in a small patch of forest located in the Gombe Stream Chimpanzee Reserve, located alongside Lake Tanganyika in what was then known as the Tanganyika Territory (modern-day Tanzania). The country had once been part of German East Africa, but at the end of World War I the region was turned over to the British with the understanding that Tanganyika would eventually be self-governing. Until that time arrived, the British acted as the presiding government for the country.

Leakey faced another obstacle: getting permission for Goodall to carry out the study. He corresponded with Geoffrey Browning, the British district commissioner of Tanganyika's Kigoma, the capital city located in the Western Province. Browning's initial responses were vague and evasive. He agreed that it might be possible for a young woman to go to the Gombe Stream Game Reserve and that she be allowed to study the chimpanzees. But Browning also told Leakey that Goodall would not be allowed under any circumstances to enter the region alone. The apes and chimps at the Reserve could be dangerous; Browning would allow Goodall to travel to the reserve only if she had an assistant who was a recognized member of her expedition.

Leakey became even more concerned: it was imperative that Goodall be paired with someone who would not threaten the project's success or compete with Goodall. The person also needed to be someone Goodall could relax around and who would not interfere with her study or the manner in which she carried it out. One day, while Goodall was out of the office,

Leakey mentioned the problem to Goodall's mother, Vanne. There passed a few moments of silence. Then, as Vanne recalled years later, she spoke up and told Leakey that she would be happy to accompany her daughter on her field study. When Leakey took Vanne's proposal to Browning, he agreed: two women, not one, would be traveling to Gombe.

Even after Leakey received official notice from the government that Jane and her companion would be allowed to go to Gombe, he met with derision and skepticism over his proposed project. Leakey was still considered a maverick in the field of paleontology, particularly in the debate over where and how exactly man evolved. What made it even more unsettling was that Leakey had clearly picked in Jane a protégée who appeared to be as much of an individualist as he.

Other academics in the field found Leakey's proposal not only strange but downright bizarre, particularly because he was entrusting the project's success to a 26-year-old woman who had once worked as a waitress and a secretary. Not only did she lack formal training, many believed this delicate-looking woman was completely incapable of surviving in the jungle for a week, much less six months. Even her enthusiasm and willingness to learn did little to silence Leakey's critics. As Goodall recalled years later, "There seriously were people who told Louis that he was practically insane."[1] But others close to Leakey recognized his genius and his willingness to take risks; he had been doing it all his life. So if anyone could pull off a six-month study of chimpanzees using an untrained young woman, Leakey was the man.

PREPARATION AND STUDY

In December1958, Goodall and her mother left Kenya for Bournemouth and the Birches. From there, Goodall went to London to begin her studies. After the two departed, Leakey received formal permission for Goodall to begin her field studies at Gombe. All that was left was to find money to underwrite the project.

An American tool manufacturer named Leighton Wilkie came to the rescue. Fascinated by Leakey's excavations of prehistoric tools and artifacts, Wilkie had funded several of Leakey's earlier projects and was more than willing to put up the money for Goodall's field study. His donation of $3,000 (approximately $19,000 in American dollars in 2004) included enough money to pay for a small boat, a tent, airfare, and any other costs that Goodall incurred during her six months of field study. As far as Goodall's lack of credentials, Wilkie was unfazed. He trusted Leakey and if Leakey said Goodall was capable of taking on the study, who was he to disagree? Still, it would be more than a year before Goodall could get underway.

After Christmas, Goodall moved into her father's home in Kensington Gardens, London. Her younger sister, Judy, a student at Guildhall School of Music, was already living with their father. To get around town, Goodall bought a cheap car that she named "Fifi." The two sisters, often accompanied by friends, frequented the "beat" coffeehouses where artists, poets, and musicians congregated. The *Troubador* was a particular favorite for the Goodall sisters. It was here that Goodall encountered her next romantic interest: a young actor by the name of Robert Young.

Despite a busy social life, Goodall tackled her studies vigorously. Two close friends of Leakey's, Osmond Hill of the London Zoo and John Napier at the Royal Free Hospital, took Goodall under their wings. She studied primate behavior with Hill and primate anatomy with Napier. To earn money, Goodall also began working at the film library of Granada Television with the Zoo Television and Film Unit, headquartered at the London Zoo in Regent's Park. When not studying, Goodall also spent a great deal of time with the animals. In addition, she worked as an assistant to Ramona Morris, the wife of the distinguished curator of mammals, Desmond Morris, who later wrote groundbreaking studies of the behavior of apes and their connections to modern man.

During the course of Jane's studies, she learned all about the early studies of chimpanzees. While she recognized the valuable contributions of those who had undertaken serious inquiry, she was horrified at two studies in which the scientists went into the jungle to collect data, which included the killing of the monkeys to authenticate the primates' ages, sexes, and ability to reproduce. For Goodall, it was just another example of how the weak and defenseless suffered because of human cruelty.

She also had the opportunity to watch the three chimpanzees at the London Zoo. There were two females and one male called Dick. Goodall later wrote that the chimps were kept in horrible conditions, particularly Dick who had been crammed into a tiny cement cage. The effect had clearly caused detrimental effects on the chimp's behavior. Dick had been shut up in the cage for so long that he was almost mad. Goodall later described Dick's behavior as consisting of little more than sitting in a corner and trying to count his fingers while his mouth continually opened and shut.

As she watched, she vowed there would come a time when she would help chimps and other primates who were exploited and mistreated. True to her word, Goodall later made that her mission. For the time being, though, she concentrated on learning all she could about the primate known as the chimpanzee.

CASTING NEW LIGHT INTO THE DARK

The first mention of the word "ape" came from the Greek philosopher Aristotle, more than three centuries before the birth of Christ. But it would be centuries more before the word "chimpanzee" was used. Before any real scientific knowledge was acquired, there appeared myths and legends of all kinds describing the strange half-man, half-beast pygmy living in the forests and jungles.

Europeans first learned about chimpanzees in 1640, when a chimp was brought to the court of the Dutch monarch, Prince William of Orange. A Dutch physicist and anatomist, Nicolaas Tulp, curious about the animal, examined the chimp. Clearly the animal did not match earlier descriptions of chimpanzees, which described the creatures as possessing both horns and hooves. A year later, Tulp wrote sardonically, "The coming of this Indian Satyr, perchance becoming famous, may dispense their [people's] dense fog."[2]

Tulp's description did little to advance people's attitudes about chimpanzees. Still, chimps continued to arrive in different countries, such as France and England, and slowly became objects of interest to other anatomists, scientists, and doctors. Fifty years after Tulp's writings, a London physician, Edward Tyson, conducted his own study of a chimp. In 1699, Tyson dissected a chimpanzee and recorded his findings. Tyson found evidence that the chimp might be related to the reputed pygmy, a race of Central Africans. The pygmies were an unusually short people, with the adults ranging from less than four to about five feet in stature. Timid and shy, pygmies dwelt in the equatorial forests. Tyson also suggested that the chimpanzee might be related to modern man, but stopped short: this "Creature so very remarkable ... [was] ... a sort of *Animal* so much resembling *Man*, that both the Ancients and Moderns have reputed it to be a *Puny Race* of Mankind."[3]

The earliest use of the name "chimpanzee" appears in the September 1738 edition of the *London Magazine*, a popular periodical of the time. In the article, the reporter wrote that "A most surprising creature is brought over ... that was taken in a wood in Guinea. She is the female of the creature which the Angolans call chimpanzee, or the mockman."[4]

Early studies of chimpanzees continued to arouse the curiosity of Europeans. Over the next two centuries, various zoological gardens (a forerunner of today's modern zoos) acquired chimpanzees. The animals proved a great source of entertainment to the zoos' visitors, causing one writer to note in 1739 that the chimps were "very pretty Company at the Tea-Table."[5] However, the animals were primarily seen as something to watch and laugh at, and little was done to provide a healthy and caring environment for them. As a result, many of the chimps, unused to the European climates, succumbed to respiratory diseases. However, in one instance, the illness of one young chimp so horrified some people that they took him from the zoo and moved him to more luxurious quarters at a local hotel where he could be properly cared for. Their actions proved too late, and the chimp's champions gathered around his bed as he lay dying. As one observer wrote later:,"He was conscious to the end, and knew his friends, but although he had no parting words to say he expressed his feelings by looks and pressure of his hand."[6]

ORIGIN OF SPECIES

More than a century later, the publication of Charles Darwin's *On The Origin of Species by Means of Natural Selection, or The Preservation of Favoured Races in the Struggle for Life* in 1859 sparked a new interest in animal behavior. Like several naturalists before him, Darwin believed that all the life on earth evolved, or developed gradually, over millions of years from a few common ancestors. It was a monumental breakthrough for science and, at the time, one of the most controversial theories about human evolution ever offered.

Fascinated with animal and plant life, Darwin, a young man of 22, served as a naturalist aboard the *H.M.S. Beagle* on a British science expedition around the world. For the next five years, Darwin found fossils of extinct animals that were similar to modern species. His greatest finds came with a visit to the Galápagos Islands, located west of Ecuador in the Pacific Ocean. While there, Darwin recorded many variations of plants and animals. He discovered that these species were similar to those he earlier found in other areas of South America. As the *H.M.S. Beagle* sailed around the world, Darwin continued his studies of plant and animal life, collecting specimens for further study.

Upon his return to London in 1836, Darwin next conducted thorough research of his notes and specimens. His studies offered several groundbreaking and controversial theories about the evolution of plants and animals. Among his theories was that evolution did occur, but that the change was a gradual process, requiring thousands to millions of years. Another theory was that the primary toll for evolution was a process called natural selection, in which the stronger of any species survived over the weaker. Darwin also proposed that the millions of species alive in the world came from a single original life-form, then evolved through a branching or dividing process called "specialization." This process eventually led to the defining characteristics of modern life. Darwin also believed that the variation within species was a random occurrence. If this were true, then the survival or extinction of any species was determined by the ability to adapt to the environment.

THE HISTORY OF PRIMATES

Darwin's theory provided an explanation for the resemblance among the various species of primates, particularly the Great Apes. Darwin believed that humans were little more than a branch on the evolutionary tree, sharing a common ancestor with another primate, the chimpanzee. Over time, fossil evidence and studies of primates offered further insight as to how primates might have evolved into a distinct species.

Among the evidence, scientists found that primates have the capability to inhabit diverse ecological areas. They also developed prehensile hands and feet capable of skillful, precise motor control. This made it possible for primates to move food from hand to mouth and perform other behaviors such as leaping and jumping from branch to branch among tall trees. Primate anatomy passed on to the mammalian skeleton included a clavicle, or collar bone, which increased mobility in the shoulder; two separated bones in the arms and legs, which also helped movement; five-digit hands and feet; nails instead of claws; and fewer kinds of teeth. Over time, primates' sense of vision improved; eyes were located at the front of the skull and color vision became more important as a means of gathering information about the environment. All of these biological adaptations were necessary for primates to move more quickly through trees and to depend less on their sense of smell, which resulted in smaller snouts.

In time, the primate brain grew larger in relation to body size. This change provided more complex behavior, enabling primates to live in larger social groups in a variety of environments. Another discovery was that primates were heavily dependant on learned behavior, sharing knowledge through social groups composed of infants, children, and adults. Complex social behaviors were learned from play, trial-and-error, problem solving, and observation.

IN THE WAKE OF DARWIN

Darwin's book sparked new interest in animal behavioral studies. Although people were interested in the behavior of monkeys and apes, most early scientific investigations focused on the anatomy of these creatures. The study of primatology, or the study of anatomy and the paleontological record of primate evolution, served as a starting point for further scientific studies.

In 1863, Thomas Henry Huxley's *Man's Place in Nature* attempted to apply Darwinism to understand the origins of humanity. In Germany, Ernst Haeckel produced an encyclopedia of primate anatomy and drew the first scientific phylogenetic trees. Since we knew the current products of human evolution, contemporary primates were seen as windows into our past and sources of understanding that could "flesh out" the fossil bones of paleontology. Anatomy was the primary focus until after 1900.

But while many scientists saw Darwin's work as a serious attempt to unlock the mysteries of evolution, *The Origin of Species* also initiated a series of bizarre theories concerning the relationship between man and monkey. Certainly one of the more curious was the idea that if man was truly descended from apes, then it would be reasonable to assume that man's early descendants be endowed with the same kinds of human qualities such as intelligence and nobility as well as a pleasant demeanor. Thirty years after Darwin's book, another Englishman, R.L. Garner, traveled to the jungles of West Africa. There, he barricaded himself into a sturdy cage, where he watched the behavior of chimpanzees. Garner also included in his studies the observation of chimpanzees in captivity. The result was unreliable and open to doubt based on Garner's wildly exaggerated claims of the extraordinary mental capabilities of the chimps.

Unfortunately, Garner's studies led to other equally dubious scientific research. In 1896, the Frenchman Victor Meunier offered a proposal for the domestication of apes and monkeys. Meunier had created an elaborate plan under which apes would be trained to perform various menial tasks such as cleaning. Further, Meunier reasoned, the apes would be more than simple machines performing simple tasks; their close relationship with man suggested that apes would be a combination "machine and mechanic."[7] One of the tasks that Meunier believed the apes would do particularly well was serving food, even announcing in some manner that "Food is Ready."[8] Meunier even saw the creation of special schools that would train the apes to be "gardeners, nursemaids, valets, chambermaids, cobblers, sailors, construction workers, painters, guards and so on."[9]

A STUDY ADVANCES

In 1912, a new era in the study of apes and chimpanzees had dawned when the Prussian Academy of Sciences established an anthropoid study station at Tenerife in the Canary Islands. Here, noted psychologist and philosopher Wolfgang Kohler carried out one of the first useful scientific studies of chimpanzees. One of Kohler's main objectives was the study of the intellectual capabilities of both man and chimp and how the two species differed in thought processes. Kohler's work was among the most scientific and honest ever done in the field. It also displayed the great affection and understanding Kohler had for his subjects.

In studying the chimpanzees, Kohler carried out a variety of tests. One included suspending bananas from the roof of the station or situating them outside the bars of the station. In each case, the chimps were capable of constructing and using tools to get the bananas. Kohler's studies at Tenerife culminated in his 1925 work, *The Mentality of Apes*, in which he concluded that not only were chimps capable of problem solving but that their behavior suggested more intensive studies were in order.

At the same time Kohler's studies were taking place at Tenerife, in Russia, a Moscow psychologist, Nadie Kohts, acquired a male chimpanzee. The chimp, named Ioni, was estimated to be about one to one-and-a-half years old. For the next two years, Kohts did an extensive study of Ioni's behavior, especially his visual perceptions. During the course of her work, Kohts wrote several papers that showed for the first time that chimps possessed color vision by having Ioni match colored cloth swatches. Kohts also found her chimp could match shapes and sizes. Moreover, Kohts reported the chimp could match a covered shape with a similar visible object, a behavior that many believed to be difficult if not impossible for nonhuman primates. Kohts concluded that Ioni's perception of the world was not markedly different from that of her own son. Ioni did not live long, and died in Koht's home at the age of five. Kohts would later repeat the same tests on her own four-year-old son and compare the results.

At the time of Kohler's published studies in 1925, another psychologist, American Robert M. Yerkes, was getting ready to conduct his own studies. After purchasing two chimps from a sailor in Boston, Yerkes took the animals to start a chimpanzee colony at the Yale Laboratories of Primate Biology. Later, Yerkes and his chimps moved to Orange Park Florida and the Yerkes Primate Laboratory (which moved again to Atlanta, Georgia as the Yerkes Regional Primate Research Center).

Yerkes's work broke much new ground in the field of primate studies. Along with his colleagues, Yerkes gathered strong scientific evidence about chimpanzees. But Yerkes wasn't completely satisfied; he now wanted to move the chimps from the lab setting to a less controlled environment. So, in 1931, a psychologist from the Yerkes Center, Winthrop Kellogg, and his wife took into their home a young chimpanzee named Gua.

For the next nine months, the Kelloggs recorded Gua's daily activities along with those of their son Donald, who was the same age as the chimp. As part of their study, the Kelloggs tested both Gua and Donald using various experimental and preschool readiness tests. One test had Gua and Donald ranking four tastes in order of preference. For Donald, this meant sweet, sour, salt, bitter. For Gua, the rankings went sour, sweet, salt, and bitter.

The Kelloggs also noted that in the beginning weeks of their study, both Donald and Gua performed about equally. Over the remaining months, though, Donald pulled ahead. The Kelloggs noted that Gua displayed considerable nonverbal communication and a certain degree of comprehension of the English language. But much to the Kelloggs' disappointment, Gua never developed any pattern of articulate speech.

Yerkes was clearly encouraged by the Kelloggs' experiments but still wanted to take his studies further. As a result, two of his students attempted unsuccessful field studies of apes. Harold Bingham, in 1929, went to Africa to study gorillas, while in 1930 Henry W. Nissen set off for the French Guinea. For the next two-and-a-half months Nissen studied chimps' behavior in their natural habitat. Unfortunately, his method of field study included traipsing around the bush with a number of porters who handled his belongings and equipment. There was little surprise, then, when the chimps, hearing the crashing about among the trees, fled from Nissen and his crew. A comparative psychologist, Nissen, however, became the first person to conduct experimental studies of the chimpanzee. Beginning in the 1920s, he performed a number of experiments on sensory deprivation using chimps as his subjects.

But not until Clarence Ray Carpenter, another of Yerkes's students, began his study of the howler monkey of Barro Colorado Island in the Panama Canal Zone in 1931 was any real progress made in field study work. Carpenter's work marked the first successful naturalistic study that also created the model for modern fieldwork, such as Jane Goodall would undertake 30 years later. Unfortunately, with the coming of World War II, scientific studies of chimpanzees were ended for the time being.

UNDERSTANDING THE "MOCK MAN"

Over the course of almost 50 years, scientists such as Kohler and Yerkes made substantial strides in the study of chimpanzees, particularly their thought processes. Kohler performed a variety of tests on his chimps to understand their visual world. In his 1925 work, *The Mentality of Apes*, he noted that "Chimpanzees manifest intelligent behavior of the general kind found in human beings . . . a type of behavior which counts as specifically human."[10] For Yerkes, the study of chimp behavior offered even more promising finds. Writing in 1943, Yerkes noted: "The study of other primates may prove the most direct and economical route to profitable knowledge of ourselves, because, in them, basic mechanisms are less obscured by cultural influence. Certainly it is unwise to assume that human biology can be advanced only by the study of man himself. This could be true only if he existed as a unique organism, lacking genetic relations to other types of creatures."[11]

Clearly Yerkes was viewing his theory first hand. By 1943, he had begun conducting studies of chimpanzees, noting how they perceived relations between color, shape, and size. Yerkes tested these abilities by placing a chimp in the middle of a room. In each corner was a box of similar size and shape. All boxes were a different color. The chimp watched as his breakfast was placed in one box. With each test, a screen was put up so the chimp could not see the person changing the location of his food to another box. As the test was repeated, the chimp consistently went to the box where his breakfast had first been placed, ignoring the color cues that had earlier helped guide the chimp to his food.

When the chimp discovered the empty box, he first searched the area around the box for his food. Unable to find it, the chimp became angry, threw himself onto the ground, and cried. The experiment was repeated in different variations: sometimes the boxes were different shapes, different sizes, and different colors. The result was always the same. Over time, the chimp learned to watch for other clues to find the location of his food.

The importance of these early studies was unquestionable. Although conducted in a controlled environment such as laboratory or a home, the studies enable scientists to penetrate the chimpanzee's cognitive abilities. Even Henry Nissen in his failed field study of 1930 thought "the ultimate capacities of these apes for complex behavior will be found and measured in our laboratory experimental situation."[12] The studies not only helped unravel the mysteries of the chimpanzee, twentieth-century pioneers such as Yerkes also set the standard for how future studies would proceed. Without them, Goodall's field study might have been a terrible disappointment.

ANOTHER DECISION TO MAKE

While waiting for Leakey to bring her some good news, Goodall continued to make a life for herself in London. But an unexpected event caused her to make a difficult decision. Her romance with the young actor Robert Young became more serious; the couple by now had fallen deeply in love and in accordance with the temper of the times made plans to marry. Young even went to Goodall's father to ask for her hand in marriage. On May 13, 1960, news of the engagement between Mr. R. B. Young and Miss V. J. Morris-Goodall appeared in the society pages of the *Daily Telegram and Morning Post*.

Unfortunately for Goodall, the timing was wrong. At the same time her engagement had been published, Leakey informed Jane that her wait was over. Everything was in place: funding, permission from the British government, and the necessary permits from the Tanganyika Game Department. She could leave now for Gombe as soon as she and Vanne could get ready. For Goodall, the decision, while hard to contemplate, never seemed in doubt. Her marriage plans were postponed indefinitely; the relationship itself would collapse in a matter of months. But once the decision was made, Goodall never looked back. Almost two weeks after her engagement had been announced, Goodall and Vanne boarded a plane to Nairobi. Her girlhood dream of going to Africa and working with animals was about to come true. On May 31, 1960, Jane Goodall was bound for the jungles in Tanganyika to study chimpanzees, "Back to the forests of Africa where it all began."[13] At the age of 27, her adventures were just beginning.

NOTES

1. "Jane Goodall," *Current Biography Yearbook, 1991* (New York: H. W. Wilson Company 1992), p. 250.
2. R. M. Yerkes and A. W. Yerkes, *The Great Apes: A Study of Anthropoid Life* (New Haven, CT: Yale University Press, 1929), p. 12.
3. R. M. Yerkes and A. W. Yerkes, *The Great Apes: A Study of Anthropoid Life* (New Haven, CT: Yale University Press, 1929), pp. 14–15.
4. Jane Goodall, *The Chimpanzees of Gombe: Patterns of Behavior* (Cambridge, MA: University of Harvard Press, 1986), p. 6.
5. D. Morris and R. Morris, *Men and the Apes* (New York: McGraw-Hill, 1966), p. 101.
6. D. Morris and R. Morris, *Men and the Apes* (New York: McGraw-Hill, 1966), p. 101.
7. Jane Goodall, *The Chimpanzees of Gombe: Patterns of Behavior* (Cambridge, MA: University of Harvard Press, 1986), p. 7.
8. Jane Goodall, *The Chimpanzees of Gombe: Patterns of Behavior* (Cambridge, MA: University of Harvard Press, 1986), p. 7.
9. D. Morris and R. Morris, *Men and the Apes* (New York: McGraw-Hill, 1966), p. 101.

10. Wolfgang Kohler, *The Mentality of Apes* (London: Routledge and Kegan Paul, 1925), p. 226.

11. R.M. Yerkes, *Chimpanzees: A Laboratory Colony* (New Haven, CT: Yale University Press, 1943), p. 3.

12. H.W. Nissen, "A Field Study of the Chimpanzee," *Comp. Psych. Monographs* 8, no. 1 (1931): 103.

13. Jane Goodall, *The Chimpanzees of Gombe: Patterns of Behavior* (Cambridge, MA: University of Harvard Press, 1986), p. 42.

Chapter 5

GOMBE

On May 31, 1960, Goodall and Vanne returned to Nairobi, Kenya. Shortly after their arrival, Louis Leakey came to visit them. He had bad news; there was an unforeseen delay. According to Leakey, a dispute of some sort had arisen between fisherman camped along the shore of the Gombe Reserve. The local game ranger was trying to resolve matters, but there was no telling how long that would take. For now, though, Leakey and game officials deemed the area too unsafe for Jane and her mother to enter. To keep them occupied, Leakey sent Jane and Vanne on a train trip to Lake Victoria. Upon their arrival, Hassan Salimu, the captain of Leakey's 42-foot cabin cruiser, the *Miocene Lady*, met them and took them to Lolui Island.

A PRACTICE RUN

Lolui was a small island measuring only 9 square miles, or approximately 5,700 acres. Covering much of the island was a thick belt of dense bush; toward the middle of Lolui, the land was grass covered with a few squat trees and big rocks scattered about. For the next three weeks, "full of enchantment,"[1] Goodall went everyday to watch the vervet monkeys that roamed the grassy plain.

Vervets hailed originally from South Africa, but over time had migrated eastward. With its black face and hands, the vervet is classified as a medium- to large-sized monkey that can weigh up to 17 pounds. In East Africa, vervets live in mountainous areas to an altitude of 13,000 feet, but they do not inhabit rain forests or deserts. Their preferred habitat is the woodlands along streams, rivers, and lakes. They are diurnal, and they sleep and eat in the trees from which they seldom venture.

Every morning before sunrise, Captain Hassan rowed Jane in a small dinghy to the island. At the end of the day, Goodall signaled Hassan, who would take her back to the cruiser. Goodall and her mother dined on a simple meal of baked beans, eggs, and tinned sausages and discussed the day's events. At night, Goodall and Vanne slept aboard the boat, which was anchored close to the island. Goodall remembers how peaceful and quiet these evenings were as she fell asleep to the gentle rhythm of the rocking boat.

A CLOSE CALL

Goodall spent her days and sometimes evenings watching the vervets eat, play, and interact, carefully recording her observations in a notebook. About 10 days passed before she moved closer to the monkeys. During the hot afternoons, she traced their movements to the interior forests where they could be closer to the lake.

At the end of two weeks, the monkeys appeared to be used to Goodall's presence. So, one afternoon she decided to follow them deeper into the forest. Moving cautiously through a tunnel of thick undergrowth, she heard unfamiliar sounds. At first she thought that some animal had scented her and was possibly frightened enough to charge. She moved deeper into the bush to avoid detection. But as she saw what was approaching, she became even more frightened. The sound she had heard came not from an animal, but from a man clad only in a loincloth and carrying a spear in one hand. Goodall thought that the man might be one of the many crocodile poachers who roamed the island. She was also sure that if she had spotted him, he no doubt had seen her.

To avoid trouble, Goodall stepped into a nearby tunnel that a hippopotamus had made for grazing and said, "Jambo—How are you?"[2] The man stopped and raised his spear, but quickly lowered it. Goodall allowed herself to relax a little, until the man began speaking in an angry voice. Although she did not understand what he was saying, Jane sensed she was in danger and deduced that the man was telling her she would be killed if he spotted her on the island again. For her own good, she should leave Lolui and never come back. With that, the man turned and walked away.

Goodall did not feel much like watching the monkeys anymore. She went to the edge of the lake and signaled for Hassan. As they returned to the boat, she told him what had happened. Upon hearing her story, Hassan became enraged and immediately rowed around the island to the spot where he knew the poachers made their camp. He spoke to them about Goodall. After several minutes, the poachers agreed to leave Goodall alone as long as she did not come to their side of the island. Goodall accepted these arrangements, but did not completely trust the poachers to live up to their part of the bargain. From then on, she remained on her guard whenever she visited Lolui.

Finally, on June 30, 1960, Goodall received welcome news. That evening a radio message from Leakey informed her that she could leave for Gombe as soon as possible. As excited as she was at last to start for Gombe, Goodall was sorry to leave the vervets. She had just become familiar with them, even naming a few: a female named Lotus, her tiny baby Grock, born only days earlier, and Pierre and Maggie. The day before their departure, both Jane and Vanne went to the island to say their goodbyes to the vervets. About to begin her field study, Goodall departed Lolui feeling as if she had left a job unfinished. But the experience had been immensely helpful in preparing her for the work ahead. Looking back on the period at the island, Goodall recalled: "The short study taught me a good deal about such things as note-taking in the field, the sort of clothes to wear, the movements a wild monkey will tolerate in a human observer and those it will not."[3]

In early July, Jane and Vanne were back in Nairobi waiting for last minute preparations to be completed before their departure. Jane spent her time shopping, visiting old friends, and trying to develop photos she had taken at Lolui. She also showed her field notes to Leakey and waited nervously while he read through them. He praised Goodall's diligence and eye for detail, which only confirmed for him that he had chosen the right person for the work at hand.

ON TO GOMBE

Before arriving at Gombe, Goodall and Vanne were to travel to Kigoma. The journey was daunting; Kigoma was more than eight hundred miles from Nairobi. Accompanying the women as far as Kigoma was Bernard Verdcourt, a botanist at the National Museum. On July 5, the threesome, along with the young boy who was to help set up camp, started out. They traveled in a Land Rover, a vehicle similar to a jeep or a Sports Utility Vehicle (SUV) that could handle the rough roads and terrain of the countryside. The Rover was packed so tightly with equipment and supplies that there was little space in which to move. Because of the overload, the vehicle swayed dangerously back and forth if driven too fast. In addition to their hazardous driving conditions, Goodall and her companions had to fight off the tsetse flies that followed their car. When they stopped along the way, swarms of the insects attacked and bit them. The flies were hard to kill because of their ability to fly away very quickly, and their bite was painful. But the tsetse posed an even greater danger: in many cases they infected their victims with "sleeping sickness," a disease that led to chronic weakness, the effects of which could last for years. If untreated, the disease could be fatal.

Despite the insects and three minor breakdowns along the way, the party arrived in Kigoma three days after starting out. Another problem awaited them. Kigoma was in chaos: only 25 miles away, on the other side of Lake Tanganyika, violent riots had broken out in the Congo.

WAITING OUT THE DELAY

Exploited and plundered by the Leopold II, the King of Belgium, who had made the region his personal possession, the Congo had a troubled past. For almost 25 years, Leopold made a sizable fortune from the production of rubber, but in the process had brutalized and killed hundreds of thousands of Congolese. Leopold's actions did not escape the attention of the international community, which mounted a number of protests. Finally, in 1908, international pressure forced the Belgian parliament to wrest control of the Congo from Leopold and adopt it as a Belgian colony. Despite the change of regime, the region, now known as the Belgian Congo, did not yet enjoy independence and was subject to continued exploitation.

On June 30, 1960, the beleaguered colony at last received its independence from Belgium. Unfortunately, the transition to freedom took a violent turn. Just days after independence had been granted, the Congolese army mutinied against their Belgian officers. The mutiny grew into a full-scale rebellion and continued until newly-elected President Kasavubu and the prime minister, Patrice-Emery Lumumba, replaced the Belgian officers with Africans. But then the Congolese turned their wrath on the Belgians and other whites, butchering thousands and causing a mass exodus to Kigoma.

When Goodall and her traveling companions drove down the main street of Kigoma a day or so after their arrival, they found it deserted. Most businesses were closed, and few people dared venture out of their homes. When Goodall finally located the regional commander, he told her that she could not leave for Gombe until the area was secure. The danger now, he explained, was that Tanganyikans might follow the lead of the Congolese and stage an uprising of their own. Goodall was devastated, but could do nothing. She could only wait to see what would happen next.

KIGOMA

Preparing for a lengthy stay, Jane, Vanne, and Verdcourt booked themselves into rooms at one of the two hotels in Kigoma. But as more refugees entered the city, the threesome were forced to adapt. Instead of each person having a private room, Vanne and her daughter doubled up, and Verdcout shared his room with two Belgian refugees. Goodall even offered their camp beds to the hotel's owner to make additional sleeping accommodations. Despite the crowded conditions, the refugees were grateful to have been spared sleeping on the cement floor of a large warehouse.

The residents of Kigoma also did their best to house and feed the refugees. Jane, Vanne, and Verdcourt volunteered to help. On their second evening in Kigoma, they, along with other volunteers, threw their energies into making two thousand Spam sandwiches and serving soup, fruit, chocolate, cigarettes, and drinks to the refugees. Afterward, Goodall said she could never look at a can of Spam again.

To make the most of their time, Goodall and her mother explored Kigoma and befriended many of the residents. The town was more like a village; the hub of activity was centered on the lakeshore where a natural harbor offered port to ships traveling back and forth from Burundi, Zambia, Malawi, and the Congo. Situated near the lake were the government offices, police headquarters, the railway station, and the post office. Goodall particularly liked visiting the local market, which offered fresh fruits and vegetables and red cooking oil made from nut palms.

As much as the two enjoyed their wanderings, Goodall was becoming anxious. Still waiting for word from local officials about when or whether she could enter Gombe, Jane grew depressed. Money was tight. Jane and her mother had to move out of the hotel. When they asked locals where they might find a suitable spot to camp, they were directed to the grounds of Kigoma prison. As it turned out, the prison grounds overlooked a lake and were beautifully landscaped. Trees bent under the weight of their fruit, lending a sweet smell to the area. Goodall's spirits lifted, although she and her mother had to contend with swarms of mosquitoes at night. Their new friends helped by providing them with meals and a place to bathe.

By this time, Goodall had all but given up hope that she would ever get to Gombe. More than a week had passed since her arrival in Kigoma, and still she had no word about the fate of her project. Just when she was thinking that she and her mother ought to return to Nairobi, David Anstey, the Game Ranger who had gone to Gombe to mediate the fishermen's dispute, returned and met with the District Commissioner. Shortly thereafter, Goodall learned that she had the necessary permission to go on to Gombe.

On July 16, 1960, Jane, her mother, their cook, Dominic, and all their equipment, including a 12-foot dinghy, were packed onto the Tanganyika Game Department's launch, the *Kibisi*. Jane remembers feeling as if she were moving in a dream. Finally, the launch moved out of Kigoma harbor and headed north toward the eastern shore of the lake. Goodall was not yet convinced that she was at last on her way, fully expecting the ship to sink or herself to fall overboard and drown or be eaten by a crocodile. Little did she realize that few believed they would ever see her again. Despite Leakey's optimism, most thought the project was doomed from the start.

ARRIVAL AT GOMBE

The Gombe Stream Game Reserve lay on a rugged, rectangular piece of land that stretches along the eastern side of Lake Tanganyika, and extendes about 10 miles from north to south. A rock and pebble shoreline formed the western boundary of the Reserve; an escarpment bordered the Reserve to the east. Approximately 15 streams flowed out of this steep eastern ledge, emptying into channels that ran through the valley below and fed the lake. The area stands in severe ecological isolation; a forest sprawled along one side of the lake, while the other side consisted of more stripped-down land, the result of overcultivation. The area was first designated a sanctuary by the German colonial government. When the British took over, they continued to maintain it as such.

In 1960, when Goodall first arrived, the reserve was still part of a larger, almost continuous wilderness habitat. The chimpanzee populations existed in the northern, southern, and eastern parts of the reserve. It was estimated that approximately 160 chimps lived in an area measuring roughly 30 square miles. In addition, the reserve was filled with other primates such as baboons, red colobus monkeys, and vervet monkeys. Hippos and buffalo once wandered the reserve, but have since disappeared, as have the crocodiles that swam along the lakeshore. Squirrels, mongooses, and other rodents were prevalent, as were many different kinds of reptiles and birds. Even endangered species such as the leopard still roamed the jungles of Gombe. For Goodall, it was as if she had entered paradise.

After about an hour, the *Kibisi* headed toward the southern boundary of the game reserve. Goodall grew nervous, wondering whether she was really up to the task ahead. How would she handle living in the wild? Where would she find chimpanzees? Would they allow her to study them? Would she know what to do? These questions and others raced through her mind as the launch landed and she stepped for the first time onto the shores of Gombe.

The game ranger, David Anstey, who had accompanied Jane and her mother, suggested that they make camp at Kasekela, a modest outpost where the Game Rangers stopped as they patrolled the reserve. A sense of fascination now replaced Goodall's anxiety. As Vanne and Anstey set up the tents, unpacked the launch, and prepared supper, Goodall walked up the forested slope opposite the camp and soon found herself sitting on a rock, looking out over the valley. She heard baboons barking in the distance and watched while birds flew and sang overhead. The aroma of grass, ripe fruit, and rich earth was intoxicating. As sunset approached, Goodall headed back to camp for supper, convinced she had been given a glimpse of heaven. She felt a deep sense of calm as she walked back down the slope, as if she had at last discovered her purpose and her place.

Anstey stayed with Jane, Vanne, and Dominic for the next few days to help them get situated. He set up a small camp for Jane and her mother. Constructed of a few poles that supported a straw roof, Dominic's quarters were located near the makeshift kitchen. Before he left, Anstey made Goodall promise that she would not venture into the hills alone, at least not until she became further acquainted with the area. Anstey enlisted a local game ranger, Adolf, to accompany her. Another local man, Rashidi Kikwale, served as Goodall's porter. Mikidadi, the son of a local chieftain, was to act as her guide.

At first, Goodall was frustrated that Anstey seemed to be monitoring her every move, but she soon understood his reasons for doing so. Anstey knew that for Goodall to gain legitimacy with the natives, it was important for her to hire locals to assist her, even if it meant they carried out menial tasks. It would also help Goodall to make friends with the locals rather than keeping them away from her work. Anstey wanted to keep Jane and her mother safe, for he was convinced they would not survive in the wild for long. He expected to be summoned to the camp within a few weeks to escort them back to Kigoma.

THE FIRST DAYS

The morning Anstey departed, Goodall arose at 5:30, ready to begin her work. She had earlier arranged to meet with Mikidadi in a valley near the northern boundary of the preserve, where Adolf had reported seeing some chimpanzees the day before. She, Adolf, and Rashidi caught up with Anstey, who agreed to take them by boat on his way to another village. When they arrived at the meeting place they found Mikidadi, who was accompanied by several of his friends. Goodall wondered if the entourage would insist on traveling with her. To her relief, Mikidadi asked her where she wanted to go. She pointed toward the forest slopes. Startled and distressed at Goodall's request, Mikidadi told her that he was feeling ill and would not be able to accompany her that day. Goodall never saw him again.

Goodall was determined to proceed without Mikidadi, and she, Adolf, and Rashidi made their way to the Mitumba Valley. After about 20 minutes, Adolf led Goodall to the side of the valley where the thick forest made walking more difficult. At one point, she and her companions had to crawl through the underbrush. Finally, Adolf stopped under a large *msululu* tree. Known for its rich, red fruit, the *msululu* was appealing to many of the animals in the area. The ground around the tree was littered with half-eaten fruit and twigs. This was the place where Adolf had earlier sighted the chimpanzees. Not wanting to disturb any chimps that might be in the tree, Goodall signaled the two men to come away to where they could watch the tree from a distance. Rashidi pointed to a grassy clearing directly opposite the tree.

As soon as Goodall took up her position, she heard the pant-hoots of a group of chimpanzees. The pant-hoot is a loud call, so loud, in fact, that observers report hearing it over a mile away in dense forest. High-ranking adult males pant-hoot most frequently, though female chimps also pant-hoot in chorus with others. The pant-hoot is used in a variety of situations, including to signal the discovery of a food source and to call or respond to the pant-hoots of other chimpanzees. Pant-hoots generally consist of four parts: the introduction, the build-up, the climax, and the let-down. Goodall soon learned that the different parts of the pant-hoot are quite distinctive, and became adept at identifying them. She also came to recognize the pant-hoots of individual chimps, which became one means of distinguishing them.

Goodall now noticed that the calls grew louder and were accompanied by another sound: chimps drumming on tree trunks. It was Rashidi who sighted the first chimpanzee as it climbed up the tree and disappeared into the branches. A strange procession followed, as Goodall watched chimp after chimp making their way up the tree. Goodall counted 16 chimps in all, including one female that was carrying a baby chimp clutched to her stomach. Goodall's initial excitement gave way to boredom. For the next two hours, the chimps did nothing except occasionally reach out to pluck a piece of fruit from the tree. They remained silent the entire time, and after two hours, they left in an orderly procession down the tree and vanished into the forest.

For the next three days, Goodall, Adolf, and Rashidi returned to the area but learned little about the chimps, except that they enjoyed eating the fruit of the *msululu* tree. What Goodall did learn was that the nature and composition of the chimpanzee group was not static; it changed regularly and often. Sometimes the chimps traveled in large groups, but at other times they went about in twos and threes. When the chimps had finished eating, they scattered in different directions.

For the next 10 days, Goodall and her party searched the valley for chimpanzees. Despite the care they took to camouflage their movements, the chimpanzees scampered away as they approached. Goodall never got close enough to observe the chimps or record their behavior. Yet, although her fieldwork had yielded few results, during those early days in Gombe Goodall was gradually becoming used to life in the jungle. Her skin toughened, and she developed some resistance to insects. Her stamina and agility improved, and each day she could venture further into the forest or up the steep slopes. Rastifi taught her to spot animal tracks and identify the different animals that made them. His knowledge and expertise became indispensable, and Goodall realized that Anstey had done her a great favor by making her take him along.

Adolf, by contrast, was becoming a problem. He was quite lazy and more interested in eating than in tracking and studying chimpanzees. By September, Goodall employed trackers who were natives to the region to replace Adolf and Rastifi. Like Rastifi, however, they were happy to teach Goodall about the jungle. She found them immensely helpful, though they did nothing to ease her growing anxiety about her lack of progress. If she did not accomplish something soon, she feared that Leakey would cancel the project.

SUCCESS!

After three months at Gombe, Goodall felt that she had learned little about the chimpanzees and time was running out. To add to her frustration, she and Vanne had contracted malaria, even though the doctor at Kigoma had told them that malaria did not exist in the region. As a consequence of this advice, they had brought with them no medicine to combat the illness. For two weeks, Jane and her mother were confined to their beds with only Dominic to look after them. As soon as she felt well enough, Goodall resumed her work, growing ever more desperate to make some kind of a breakthrough.

One morning Goodall arose at her usual time and went off by herself. In the coolness of the morning she headed for the mountain that she had visited on her first day at Gombe. Still weak from her illness, she paced herself and rested frequently. She climbed to about 1,000 feet above the lake, from which vantage point she could view the Kakombe valley to the north and the Kasaleka valley where her camp was located. She sat down, took out the binoculars she had brought, and began to scout for chimps.

She had no need of binoculars. As she surveyed the valleys below, Goodall heard something moving about eight yards from where she was sitting. She turned around and to her surprise and delight she saw three chimpanzees staring at her. Goodall expected the chimps to run, but instead they paused before disappearing into the brush and trees.

Goodall remained at the spot throughout the morning, hoping the chimps would return. Suddenly, she heard a clamor screaming, barking, and hooting as a parade of chimps made their way to a fig tree just below where she sat. Soon another group arrived and, pausing momentarily, joined the others at the fig tree. The noise died away and the chimps became very quiet. When they had finished eating, the chimps walked away from the tree in a straight line, stopping only to drink water before going on their way. Goodall observed the chimps closely and furiously scribbled notes. The episode marked her best day's work since coming to Gombe. It was the breakthrough for which she had been hoping.

THE ROUTINE

After her experience at the Peak, as Jane now referred to the spot where she had encountered the chimps, she began to rise every morning at 5:30. She made coffee, ate a slice of bread for breakfast, and began to climb the slope while it was still dark. She took with her only a small trunk that contained a sweater, a blanket, a few tins of baked beans, and a small bowl with which to catch water from the nearby stream. Often she went without lunch, though she occasionally would eat some of the ripe fruit the chimps fed on. It was exhausting to climb so high, far, and fast. By midafternoon, Goodall was tired and dirty, having spent most of the day on her stomach crawling in the dirt, vines catching in her hair, and suffering cuts, scrapes, and bruises from climbing through the thick brush.

Sometimes she remained at the Peak overnight. In a 1999 interview, Goodall described those days waiting for the chimps to appear:

> Fortunately I'd learned to be patient in the hen house, and I found a rocky peak where I could sit, wearing the same colored clothes everyday, using my binoculars, not pushing, not trying to get too close too quickly.[4]

As the days went by, Goodall found there were few things she wanted or needed from the outside world. The one luxury she missed was listening to classical music. Goodall also looked forward to the days when Dominic and Rashidi traveled to Kigoma to pick up fresh fruit and meat. They also arrived bearing mail from England, which helped Goodall during those moments when she felt homesick for her friends and family.

Her patience and persistence paid off. Only rarely did the chimpanzees fail to come to the Peak. Jane observed them closely and, after they had departed, examined the debris they had left behind and explored the intricate nests they made in the trees by weaving together twigs and branches. Jane began to recognize patterns in the behavior of the chimpanzees. She noted, for instance, that they tended to travel in groups of six or fewer, and that each group commonly included a mother with children and between two and three adult males. Sometimes the groups split up and reformed as other chimps arrived and departed. In time, some of the chimps appeared to be as interested in Jane as she was in them, though they remained hesitant to approach her for about a year. But they no longer ran away when they saw her as they once had.

In another interview, Goodall described the rewards of her work: "The most wonderful thing about fieldwork, whether with chimps, baboons or any other wildlife, is waking up and asking yourself, 'What am I going to see today?'"[5] By far Goodall's favorite days were spent following a mother and her family until evening. She enjoyed watching the youngsters at play and how the mother chimps lovingly interacted with their children.

Goodall eventually came to realize that the chimps she observed constituted one large group consisting of approximately 50 chimps, which she named the Kasakela group. She began to identify individuals and, instead of assigning them numbers, named them, often selecting names because the chimps reminded her of people she knew. There were Goliath, David Greybeard, Flo, her mother Olly, Mr. McGregor, and William, whom Goodall believed was Olly's brother. Each chimp had distinctive characteristics. David Greybeard, for instance, had a dense white beard and beautiful intelligent eyes. He was a handsome chimp with a gentle and calming nature. He often traveled with Goliath, so-called because of his size. He was at the peak of his strength when Goodall first began watching him. She was amazed at his power, strength, and agility.

Mr. McGregor was the easiest to spot: the chimp was bald from the crown of his head to his neck and shoulders. What little hair that remained on his neck reminded Goodall of a monk's tonsure. His was a belligerent personality; he often threatened Goodall by shaking the branches. William was the clown with a wobbly upper lip and deeply etched scars that ran from his upper lip to his nose. He was a passive personality often picked on by the other chimps.

Then there was Flo, one of the older chimps. She was not a pretty chimp; her deformed bulbous nose and ragged ears made her easy to recognize. But she was also one of the strongest females in the Kasakela community; friendly with both males and females, she had an energy and gentleness that seemed to draw many chimps to her. She was often accompanied by her daughter Fifi and her son Figan. It was Flo who would teach Goodall more about chimp sexuality and parenting than any other chimp.

MAJOR BREAKTHROUGHS

During the next few months, Goodall made two important discoveries that not only proved the study was worth continuing but that radically changed its scope and direction.

The first breakthrough came on a visit to the Peak. When Goodall arrived, she saw a group of chimps sitting in the branches of a tall tree. She recognized some of them, and watched as David Greybeard pulled something pink out of his mouth. He then plucked some leaves and chewed for several minutes before spitting the mixture out and giving it to one of the females and the baby chimp sitting next to its mother. To her surprise, Goodall realized that the chimps were eating meat. At the base of the tree, she spied the carcasses of three small piglets. From this one incident, Goodall had dispelled the conventional wisdom that chimpanzees were vegetarian.

The second event occurred two weeks later, in early November. Before Jane had left for Gombe, Leakey had instructed her to determine whether chimps could make and use tools. Thus far, she had seen no evidence of tool-making ability. Then, during one visit to the Peak, she observed David Greybeard sitting on a termite hill. Through her binoculars, she watched as David repeatedly pushed a small blade of grass into the hill. Each time he pulled the stem up he picked something from it and put it into his mouth. After David left, Goodall approached the termite hill and repeated David's actions. Each time she pulled up the stem it was covered with termites. David Greybeard had used a tool to collect food! On subsequent observations of the termite hill, Goodall noted how other chimps made small openings in the hill with their hands and then inserted a twig to gather termites. Goodall realized the significance of her findings. As far as she knew, these were the first recorded instances of animals, other than man, making and using tools.

The implications were enormous and sent shock waves through the scientific community. For years, anthropologists believed man was the only creature capable of making and using tools. But Goodall had proved otherwise. As soon as she could, Goodall telegraphed the news to Leakey. He responded with unrestrained excitement: "Ah! We must now redefine man, redefine tools, or accept the chimpanzees as human!"[6]

At the same time, many scientists were troubled by Jane's findings, for they challenged the uniqueness of humanity. Perhaps humankind was more closely related to animals than scientists cared to admit. Some scientists even tried to discredit her discovery, claiming that Jane lacked the professional training to enable her accurately to document her observations. One scientist went so far as to suggest that Goodall had staged the whole affair by training the chimps to retrieve termites using grass and twigs. Yet, there were also those who accepted Jane's initial findings and called for additional research to verify them.

HAPPY NEWS, SAD NEWS

In any event, the controversy served Leakey well, giving him all the ammunition that he needed to secure additional funding for Jane's work. Jane, of course, was elated at the prospect of remaining at Gombe, but after five months in the jungle, Vanne wanted to return to England. Vanne had not contributed much to Jane's research, but she had improved relations with the locals by administering medical aid to them. While Jane studied chimpanzees, Vanne busied herself with establishing a small clinic. With medical supplies that her brother Eric sent from England, Vanne did what she could. She distributed aspirin or Epsom salts, but in one case administered a saline drip to treat a stomach ulcer. Through her patience, aid, and eagerness to understand the locals, Vanne ensured the friendship from which her daughter would benefit in the years to come. Goodall has long credited her mother with making her work at Gombe so successful.

As Vanne prepared to leave, Goodall received a wonderful piece of news. An American organization, the National Geographic Society, had agreed to provide the sum of $1,400 (approximately $8,700 in 2004 dollars) for Goodall to continue her field study. The society, founded in 1888, was organized to advance the general knowledge of geography and the world among the general public. To this end, it sponsored exploration trips and published a monthly magazine, *National Geographic*. For Goodall, it marked the beginning of a long relationship with the organization. The grant was the first of many that she would receive from the organization over the next several years. News of the grant could not have come at a better time, for with Vanne's impending departure, Goodall realized that for the first time in her life she would have to carry on alone. The grant would surely make that easier.

NOTES

1. Jane Goodall, *In the Shadow of Man* (Boston: Houghton Mifflin, 1971), p. 8.

2. Jane Goodall, *My Life with Chimpanzees* (New York: Simon and Schuster, 1996), p. 53.

3. Jane Goodall, *In the Shadow of Man* (Boston: Houghton Mifflin, 1971), p. 8.

4. Michael Toms, "Born to Be Wild: A Conversation with Jane Goodall," New Dimensions, World Broadcasting Network, 1999, http://www.newdimensions.org/online-journal/articles/born-to-be-wild.html (accessed June 10, 2004).

5. "A Day in the Life of Jane Goodall," Jane Goodall Institute, http://www.janegoodall.ca/jane/jane_bio_day.html (accessed on: September 10, 2004).

6. Jane Goodall with Phillip Berman, *Reason for Hope: A Spiritual Journey* (New York: Time-Warner, 1999), p. 67.

A young Jane Goodall at work on her study of baboon behavior at her research station in Gombe National Park. (Photofest)

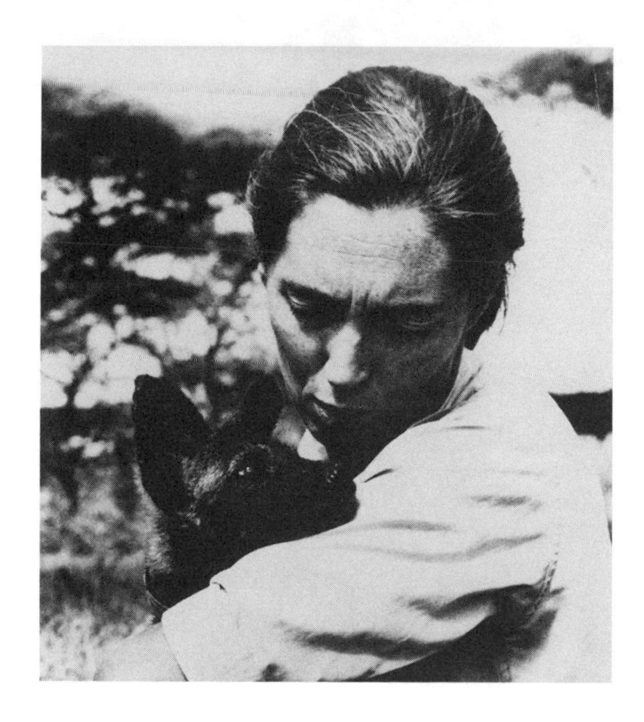

Goodall tames a pup from one of the vicious packs of wild dogs that roam Africa. (Photofest)

The legendary animal behaviorist has an unmatched understanding of chimpanzees after more than 30 years of study. (Photofest)

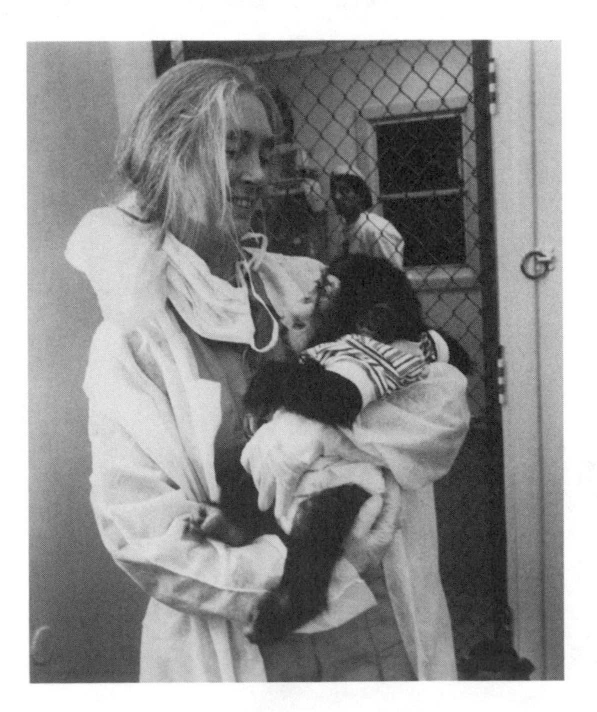

Jane Goodall—visiting a research lab in the United States. (Photofest)

Speaking during an interview. (Photofest)

During one of her many observations of chimpanzees. (Photofest)

Chapter 6
THE BANANA CLUB

As it turned out, Jane was not alone for long after her mother's departure. Dominic brought his wife and daughter to live with him in the camp, and Leakey sent Hassan with his wife and child. To build on the goodwill that Vanne had created, Goodall learned Swahili and assimilated to the native customs and culture. As a result, locals frequented the camp. At the same time, a growing menagerie came to call it home.

In addition, Goodall regularly entertained guests. David Anstey visited on occasion, as did friends from Kigoma. Among the most noteworthy visitors were the zoologist George Schaller and his wife. Schaller had recently completed a pioneering study of gorillas in Rwanda. He told Goodall that if she could prove chimps ate meat and made tools, her work would make an enduring contribution to science.

BECOMING ONE WITH THE LAND

At the end of November, Goodall returned to Kenya for six weeks, but was back in Gombe by mid-January and hard at work. Although she missed her mother, Jane did not feel isolated or lonely. She was more than ever at home in her surroundings. Rainy days were a particular favorite. Without fail, Jane went to the jungle and there sat in silence for hours listening to the rain patter against the leaves and the ground, hidden away from the world.

The rainy season, though, forced Jane to alter her methods. She kept her binoculars dry by wrapping them in a tube of plastic. Still, moisture formed on the lenses. Although Jane wore a poncho when it rained, she tended to become drenched and cold. To keep her clothes as dry as possible, she took to walking to the Peak naked on rainy days, carrying her clothes with her in a small plastic bag.

THE CHIMPANZEE RAIN DANCE

Some thought Goodall's findings were the result of extraordinary good luck. Those who knew her understood that "luck" had nothing to do with Jane's success. It was, rather, the result of persistent hard work. In an interview, Jane explained why she believed she had succeed in studying and understanding chimpanzees when others had failed: "I've always felt you don't have to be completely detached, emotionally uninvolved, to make precise observations. There's nothing wrong with feeling great empathy for your subjects."[1] Jane's empathy for the chimps deepened during the rainy season when she caught glimpses of them huddled together beneath trees, cold, wet, and miserable. She noticed that baby chimps fared much better in bad weather because the mothers protected them from the elements. Goodall felt guilty going back to camp, where hot food and a warm bed awaited her.

But watching the chimps in a rainstorm also provided Goodall with another important insight into their nature: their playfulness. Once as she saw a storm approaching, she noticed a group of chimps in a nearby tree. When the rain began to fall, the male chimps climbed down and headed for a ridge. With the first clap of thunder and streak of lightning, a male chimp began to pant-hoot and swagger back and forth. Then he ran about 30 yards down the slope toward the trees the group had just left. He swung around the trunk of a small tree and leaped into the branches remaining perfectly still. Two other males then charged down the slope. One broke off a branch and hurled it. The other stood upright and swayed the branches of a tree back and forth, then breaking off a large branch, dragged it further down the slope. As Goodall watched in amazement from beneath a plastic sheet, the chimp that had begun the presentation jumped down from the tree and ran up the slope. The others followed, plodding their way to the top. When they had reached the open ridge, they charged down the slope one by one with as much energy and enthusiasm as before, repeating their earlier performance. All the while, the female chimps and children watched from the nearby trees. After about 20 minutes, the male chimps ended their frolic. The females and children climbed down and the group disappeared over the crest of the ridge. As the last male passed Jane, he paused, placing his hand on the trunk of tree. She likened the scene to an actor taking a final bow before the curtain fell. Then that chimp, too, lumbered over the ridge to join the others.

For Goodall, viewing the chimpanzee rain dance was a riveting experience. She marveled at the elegance, grace, and strength that the chimps possessed, and wondered whether primitive men might not have similarly entertained themselves, defied the elements, and tried, if only briefly, to become one with the gods.

A CLOSE ENCOUNTER

During the next several months, Goodall continued to watch the chimpanzees. She learned that she could often move closer to the chimps when the weather was cold or rainy, than when warmer, sunnier conditions prevailed. On one such rainy day, she walked through the forest silently trying to catch up with a group of chimps she had heard before the rain had started. As she spotted a black shape a few yards ahead of her hunched on the ground, Goodall lay low hoping that the chimp had not caught sight her. She turned away for a moment when she heard a slight rustle among the leaves. When she turned back, the chimp was gone.

Before she had a chance to feel discouraged, Goodall heard another sound coming from above the spot where she lay. Looking up, she recognized Goliath staring down at her, his lips tensed. He shook a branch at Goodall, who turned away not wanting to pose a threat. As she did so, she saw another chimp, its arm tightly gripping a vine. More rustling told her that a chimp had come up behind her. She was surrounded.

Her first impulse was to run, but where could she go? Instead, she stood still and waited. Goliath let out a yell and, shaking the tree limb, showered Goodall with all manner of twigs and branches. The two other chimps also began to yell; Goodall described the racket as one of the most savage sounds she had ever heard. Although frightened, she did not move from her place. Then she felt something strike her. It was the end of a branch. One of the chimps ran toward her, but veered into the bush.

Goodall wrote that she expected to be torn apart. After the chimps had fled, she remained in her spot for a long time, making sure she was out of danger. Her knees shook as she stood up, but she felt a mixture of relief and exhilaration at having escaped danger. She also realized her relation with the chimps had changed. They were not as fearful of her as they once had been.

From this moment, and for the next five months, the chimpanzees displayed more aggression and hostility toward Goodall. Only weeks earlier, they had hidden themselves whenever they saw or heard her approach. Now, they challenged and menaced her.

Another startling incident occurred about three weeks after the encounter with Goliath. Goodall had positioned herself along a narrow ravine, hoping that some chimps might come to eat from a nearby tree directly opposite the spot where she waited. Soon she heard the unmistakable sound of footsteps coming up behind her. She flattened herself out on the slope and waited, not making a movement or a sound. Then Goodall heard a soft cry and the footsteps had stopped, close to where she lay. Silence followed. Again Goodall heard the footsteps coming closer followed by a scream so loud and unearthly that it gave Goodall a start. She turned around in time to see a large male chimp scampering up a tree just above her head. The chimp shook the branches and continued screaming at Goodall. He jumped down and began striking the trunk, working himself into a frenzy. He broke off a low branch and came toward Goodall, menacingly showing his yellow teeth and pink tongue. He shook the branch at Goodall, covering her with twigs. Then, just as quickly, he stopped and walked away.

When she had composed herself, Goodall noticed a female chimp sitting in a nearby tree holding a tiny baby. An older child sat next to her watching Goodall, its eyes open wide. Goodall heard the footsteps of the male again coming up behind her. He came so close that she could hear him breathing. Without warning, he struck her hard in the head. She fell, but sat up quickly and saw the male chimp staring down at her. She was afraid that he might charge, but instead he again turned around and walked away, glancing back at her over his shoulder every so often. The female chimp and her two children climbed down from their perch and raced to catch up with the male. Goodall was alone again. Relieved to have survived, she realized that this incident counted as her first real contact with the chimps at Gombe, though she admitted that she did not know whether it was she who made the contact or whether it was the chimp.

Some time later, Goodall was able to make sense of what took place that day. The chimp that had struck her was probably the chimp she called J. B. He had a terrible temper and an irritable personality. Goodall believed that he simply wanted to know what she was and could not tell because of the plastic poncho she wore that day. J. B. knew she was alive, because he had seen Goodall's eyes. But he needed to make her move and determine whether she was a threat to him.

When Goodall returned to camp she told Hassan and Dominic what had happened. The two passed on the story to Iddi Matata, the chief of a nearby village. That evening the chief came to visit Goodall and told her the story of a man who had one day climbed a palm tree to pick fruit. Unbeknownst to him, a male chimpanzee was already in the tree. Before the man realized what had happened, the chimp attacked the man, hitting him in the face before scampering down the trunk and running away. As a result of the blow, the man lost sight in one eye. The chief told Jane that many people of Gombe believed that she had magic. Wherever she went, she suffered no harm, though others had been hurt or killed under similar circumstances. The belief in Jane's special powers elevated her stature among, and strengthened her relations with, her African neighbors.

A DIFFICULT SUMMER

In June, the rains subsided though the days remained gray and cold. As the season gradually turned drier, warmer, and sunnier, the jungle began to feel like a giant greenhouse in which the air hung heavy and still. Jane had been ill since May, so weakened by fever that at times she had barely the strength to raise her hand. When the humidity of the jungle enveloped her, breathing became a chore and she sometimes climbed a tree so that she could breathe more easily, though in her condition the effort exhausted her.

Although illness severely restricted her activity, Goodall experienced two incidents that convinced her she was making progress. The first happened when Goodall, having ventured to the Peak, noticed a group of chimps in the forest below. The chimps had gathered around a fig tree and were eating peacefully. To get a better view, Goodall slowly moved down the slope, hoping to conceal herself behind a large tree. As she moved cautiously around the tree, her heart sank: the chimps had disappeared. Discouraged and tired, Goodall decided to return to camp. As she turned to go she was surprised to see two male chimps sitting on the ground 20 yards from her. They stared at her intensely. Goodall stopped moving. Minutes passed and the two chimps continued to look at her. Goodall sat down, and the two chimps, taking no further notice of her, began grooming each other. She recognized David Greybeard and Goliath, whom she came to understood were high-ranking males in the community.

Then two more chimps peered over the tall grasses: a female and her young child. When Goodall turned to look at them, they ducked into the grass, but reappeared beneath a tree about 40 yards away. They, too, sat quietly watching Goodall.

Goodall continued to watch Goliath and David Greybeard groom each other for another 10 minutes. Then, as the sun began to set, David Greybeard stood up. Goodall followed suit, and her elongated shadow fell across David Greybeard's face. He looked into Goodall's eyes. It was a powerful moment, ripe with significance. The chimpanzees, Jane thought, were at last coming to accept her presence.

The second incident occurred a few weeks later. By now, Goodall was spending less time on the Peak. She had instead moved deeper into the valley to get closer to the chimps. One day, while watching a group of chimps from about 30 yards away, she heard a rustle in the leaves nearby. She looked around and tried to conceal her excitement. Less than 15 feet away sat a chimpanzee with his back to her. She remained still, not wishing to disturb or frighten him. After a few moments, the chimp turned around and, seeing Jane, resumed eating. When he had finished, he stood up and walked away. Goodall recognized the chimp as Mike, an adult male almost as handsome as David Greybeard. This incident represented another milestone. Fear had given way to aggression and aggression had now given way to acceptance.

REACHING A BROADER AUDIENCE

While Jane was chiefly interested, first, in getting the chimpanzees to stop being frightened of her, and then, to stop attacking her, others were taking a different kind of interest in her work. In the late spring of 1961, the National Geographic Society set out to document Goodall's findings, especially her claim that chimpanzees could make and use tools. An editor flew to Nairobi to interview Louis Leakey. He explained that the magazine wanted to do a story about Goodall's work and take photographs showing her with the chimps.

Before the discussion went further, Leakey pointed out the difficulties that such an undertaking involved. Besides carrying heavy, cumbersome, and expensive equipment into the jungle, the photographer would have to deal with potentially elusive subjects. Not only were the chimpanzees a dark color, but they also tended to conceal themselves from view. If the chimps did not want to be photographed, it would be next to impossible to do so.

In trying to discourage the National Geographic Society, Leakey was, of course, also concerned that the proposed article would cast Goodall's work in an unfavorable light. Yet, given the progress she had made, Leakey found it hard to refuse the request. He finally relented and told the magazine's editors that they could send a photographer in the summer of 1961. Yet, Leakey warned that, given the delicate nature of the project, Goodall feared the sudden appearance of a stranger would disrupt the fragile relationship she had built with the chimpanzees.

To accommodate the photographer without disturbing the chimps, Goodall and her companions built a series of "blinds." These small structures would allow a photographer to shoot pictures virtually unseen. But the project hit a snag when the woman chosen to shoot the pictures for the story balked at the conditions under which she had to work. The editors at *National Geographic* then suggested that Leakey send his son Richard to photograph the chimps, but Leakey vetoed the idea. Perhaps, he thought, Goodall herself could shoot the photos. The editors agreed, but sent Goodall a complicated camera that she could not figure out how to use. The affair became increasingly absurd, and in a letter dated August 14, 1961, Goodall informed *National Geographic* that she would not take any photographs.

She proposed instead that her sister Judy come to Gombe. Although Judy was not a professional photographer, she did understand her sister's work and could take the photographs in such a way that it did not interfere with it. The Society rejected the offer. But the British newspaper *Revielle* offered to pay Judy's transportation and expenses to Gombe in return for an exclusive series of interviews with Goodall.

Judy arrived in late August, but the rainy season destroyed any opportunities to take photographs. Judy spent countless hours beneath a poncho trying to stay dry and set up a shot. She had little success. Not until November was Judy able to take photos of the chimps and of her sister in camp and in the field.

A TRIP TO CAMBRIDGE

While Goodall and her sister were at Gombe, Leakey had arranged for Jane to return to England to begin her doctoral studies at Cambridge University. She planned to study under Professor Robert Hinde, noted for his work in zoology and animal behavior. It was important to Leakey that Goodall be accepted into the academic community so that her research would be taken more seriously.

By the time Goodall arrived at Cambridge, the field of ethology was undergoing a critical transformation from naturalists to scientists known as ethologists. By the late twentieth century, naturalists had been dismissed as being little more than self-taught amateurs and explorers who merely scribbled notes in their tattered field guides. By the 1960s, the old-style naturalists who conveyed their observations through narratives, sketches, and drawings had been replaced by young men in starched white lab coats, pursuing the study of primates through logic, measurements, and experiments. They were hailed as the new messiahs who would deliver humans control over their environment, which would ensure greater comfort, productivity, and affluence.

Many early naturalists believed that animals had thoughts, feelings, and motives. Konrad Lorenz, a noted Bavarian naturalist during the early twentieth century, was one of the stronger proponents of this viewpoint. Lorenz studied geese and was able to compile solid findings without ever having to conduct an experiment. Lorenz did not believe in interfering with nature; relying on his own senses, intuitive thinking, and observation, he succeeded in understanding animal behavior. It is the spirit of Lorenz with which Goodall most closely identified.

By the late 1950s, Lorenz's influence was fading; ethology had become an experimental science. It was brutally objective, quantitative, and experimental. The goal was to solve problems, not stand by and observe. The works of many early naturalists were disregarded; they were, after all, just insect collectors. If one did not have a theory or the numbers to back it up, then there was little sense in wasting time writing down everything one had seen. It was the strength of this bias that makes Goodall's research all the more remarkable.

Goodall began her studies in January 1962, but she thought endlessly about Gombe and the chimpanzees, counting the days until she could return. She found Cambridge cold and gray, and fought regularly with the leaky pipes in her room. After two months at Cambridge, Goodall prepared to go back to Africa. By that time, she had become something of an academic celebrity. She addressed two scientific conferences, one in London, the other in New York City. Her research was exciting great interest, and many scientists wanted to know more about her methods and her findings. In addition, the National Geographic Society was more convinced than ever that Goodall's remarkable story needed to be told. They asked whether Goodall would be willing to fly back to Gombe that February so that a photographer could take pictures of her. Both Leakey and Goodall refused to cooperate, stating that the proposed time frame offered too short a notice. Despite her refusal, the Society continued to support Goodall's work both at Cambridge and at Gombe. In April, the Society paid for a trip to its headquarters in Washington, D.C., so that Jane could present her findings to the Society's board of trustees. Favorably impressed by Goodall and her work, the board presented her with the 1962 Franklin L. Burr Prize for her contribution to science. The prize carried a stipend of $1,500. With ongoing financial support from the National Geographic Society, Goodall excitedly finalized her plans to return to Gombe.

DAVID GREYBEARD

While she was in Cambridge, Goodall had feared that the chimpanzees would not recognize her when she returned to Gombe. She need not have worried. If anything, they seemed more familiar and tolerant than before. One evening shortly after her return, Goodall was walking back to camp when Hassan and Dominic came out to meet her. Both were very excited. They told her that a male chimpanzee had spent nearly an hour feeding from a palm tree located directly over her tent. He returned the next day.

After breakfast the following morning, as Goodall typed her field notes, she was astonished to see David Greybeard stroll into the camp, walk past her tent, and climb the palm tree. For the next several minutes Goodall could hear only his low grunts as he ate. After about an hour, he climbed down, paused, looked deliberately into Goodall's tent, and went on his way. From then on, David Greybeard appeared at the camp every day to eat.

One day as Goodall was sitting outside her tent, David Greybeard as usual climbed down from the tree. On this occasion, though, he walked straight toward her. His hair stood on end, which gave him a very fierce look indeed. Goodall was puzzled at this behavior, for she knew chimps raised their hackles only when they were angry, frustrated, or nervous. Suddenly David charged and grabbed a banana from a nearby table. He hurried off with his treasure, and did not eat it until he was at a safe distance. After this incident, Goodall instructed Dominic to set out bananas every day.

THE BANANA CLUB

As excited as she was about David Greybeard's visits, Goodall was eager to get back into the field. But after two months, she again fell ill with malaria and was confined to bed. She told Dominic to continue putting out the bananas for David Greybeard. Late one morning, Goodall watched as David picked up some bananas from the table outside her tent. As he walked toward the bushes, Goodall saw another chimp waiting, half hidden in the grass. It was Goliath, who watched as David ate. Goliath held his hand out, no doubt hoping his friend would share part of the banana with him. David Greybeard obliged, spiting a large wad of mushy banana into Goliath's hand.

The next day, David and Goliath returned to camp again. This time Goliath hesitantly walked up to the table with David, and took some bananas for himself. Watching the scene from her bed through a small hole in her tent, Goodall was ecstatic. She then asked Dominic to go to a neighboring village and buy a large supply of bananas. Thus began what became known as the Banana Club.

Each day, Goodall or Dominic placed a large bunch of bananas on the table and waited for the chimpanzees to appear. By now, the figs were also ripe and they began attracting other chimps to the camp. Nearly recovered from her bout with malaria, Goodall spent her time observing the chimps at the fig trees and waiting for David and Goliath to come into the camp. David continued to be a regular visitor, sometimes coming alone and sometimes in the company of Goliath. On some days, David also brought William, or other young chimps, with him.

As she grew more familiar with David Greybeard, Goodall decided to try an experiment. One day when David had come into the camp alone, Goodall waited for him beside the table outside, a banana in her hand. David approached her, raised his hair and jerked his chin upward. Goodall knew that these were mildly threatening gestures. She did not move and held onto the banana. David began swaggering from one foot to the other, and slapped the trunk of the nearby palm tree. Goodall stood fast. David ended his display, reached out his hand, and very gently took the banana from Goodall.

When Goodall tried the same experiment with Goliath, she got different results. The first time Goodall offered Goliath a banana, he raised his hair, grabbed a nearby chair, and charged at Goodall, almost knocking her over. Then he turned and ran into the bushes where he sat glowering. It was a long time before Goliath took a banana from Goodall's hand.

The creation of the Banana Club enabled Goodall to observe the same chimps on a regular basis under the same conditions. Because chimpanzees generally do not follow a routine, Goodall's attempts to monitor the same chimps under the same conditions was almost impossible. But when certain chimps, especially David Greybeard, Goliath, and William, began visiting the camp more regularly, Goodall's field notes became more detailed. She was also better able to evaluate the three when she went out into the field to observe them interacting with their community.

During this time Goodall started to piece together the organization and structure of the group. As it turned out, Goliath was the supreme male chimpanzee. If William happened to be eating and Goliath moved toward the food, William backed away and let Goliath take what he wanted. If Goliath met another male chimp, that chimp stepped aside to let him pass. Goliath was also the first to be greeted when a newcomer joined the group, and he showed no qualms about taking nests away from other chimps if it suited him.

William, by contrast, was among the more passive and submissive males. If another male became aggressive, William made conciliatory gestures hoping to avoid a confrontation. William even seemed afraid of Goodall. He would not approach her and accept a banana from her hand. She instead had to place it on the ground.

By far, David Greybeard had the hardest personality to decipher. Goodall recognized that he was one of the more sedate chimpanzees in the group. If approached by a submissive chimp, David reached out with a reassuring gesture, often laying his hand on the other chimp's shoulders or head. Sometimes, David groomed the more submissive chimp. When in the company of Goliath who tended to be excitable, David calmed him by stroking or grooming him.

Goodall realized that, like people, chimpanzees were capable of a wide range of emotions. They were not, as scientists had previously thought, uniformly vicious and harmful creatures. Often, they could be, but were often quite gentle and seemed capable of loyalty, friendship, and compassion. In her first field report to Leakey, Jane's observations called into question many scientific preconceptions about chimpanzees. Her confidence in her ability and her work was growing. In one letter to Leakey, Goodall reflected that "the challenge has been met. The hills and forests are my home. And what is more, I think my mind works like a chimp's subconsciously."[2]

NOTES

1. Ron Arias, "Jane Goodall," *People Weekly*, May 14, 1990, p. 95.
2. Jane Goodall, *Africa in My Blood: An Autobiography In Letters, The Early Years*, ed. Dale Peterson (New York: Houghton Mifflin, 2000), p. 157.

Chapter 7

THE JOURNEY DEEPENS

The more time she spent observing chimpanzees, the more Goodall became convinced that they possessed some remarkable qualities. They could reason and solve problems, much as David Greybeard had done when fishing for termites. They could fashion and use simple tools, and could modify them as needed. They could also make weapons, such as the stones males threw at each other with amazing accuracy.

Goodall, however, had become most fascinated with the chimps' ability to communicate and in the broad range of their emotional responses. In these respects, perhaps, the chimpanzees were most similar to human beings. They hugged, kissed, held hands, and patted one another on the back. They also punched, kicked, pinched, and swaggered. Chimps, Goodall decided, also held grudges that sometimes lasted for weeks. Many formed strong family bonds and intimate friendships. The society of chimpanzees was far more complex than Goodall had ever imagined. Group membership was constantly changing. She watched as chimps decided whom to travel with on a given day, or made up their minds to go off with on their own. Her field notes underscored the conclusions that Wolfgang Kohler had reached almost 40 years earlier when he stated that "chimpanzees manifest intelligent behavior of the general kind familiar to human beings ... a type of behavior which counts as specifically human."[1]

DAVID GREYBEARD'S INVITATION

Goodall's interest in the chimpanzees was more than scientific; it was also spiritual. She began to feel a deep inner peace and contentment as she went about her daily routine in a land that she had now come to think of as home. When she observed chimps such as David Greybeard eating figs from a tree, she experienced a sense of oneness and harmony with nature. Goodall believed that over time she and David Greybeard had developed a relationship, perhaps even a friendship of sorts. One afternoon as she watched David eat, Goodall was disturbed from her reverie by a shower of twigs and an overripe fig that fell at her feet and splattered on the ground. She watched as David came down from the tree, moved a few paces back from her, and sat down. He groomed himself and laid back, with one hand resting underneath his head. Like Goodall, he began gazing upward. Goodall realized that he had offered her something special: the acceptance of a wild animal. Goodall regarded David's acceptance as a gift and vowed to herself never to take it for granted.

When David stood up and walked away, Goodall followed him. As he moved toward a stream, Goodall feared she would lose sight of him and struggled to keep pace. When she reached the stream, however, David sat on the bank as if waiting for her to join him. Goodall glanced into his eyes, knowing that as long as she maintained eye contact without arrogance or pleading, he accepted her presence. She noticed some fruit lying on the ground. She picked it up and offered it to David. He reached out to take it, then dropped it, but his hand still rested in Goodall's. He did not want the fruit, but understanding her intent reassured her that she had done nothing to offend him. Goodall was deeply moved. When David got up and walked away, Goodall stayed and contemplated what had just taken place. The two had communicated, not in words but in a language far older than writing or speech. It was an extraordinary moment, Goodall thought, when two worlds and two species were briefly joined in friendship.

David continued to act as Goodall's guide in the intricate world of the chimpanzees. On one occasion, Goodall followed him as he made his way toward a group of chimps. Knowing that Goodall was behind him, he whistled and pant-hooted a greeting; when the other chimps replied, he hooted again announcing his arrival. When he approached the tree, Goodall stayed behind and watched the chimps greet David. Goliath embraced him. Then David began eating with the others. After they had finished, the chimps climbed down from the tree one by one. But instead of walking away, the youngsters began playing and chasing each other, while the adults settled down to sleep or to groom. Goodall watched the scene unfold with growing amazement. When David and Goliath got up a few hours later they walked away together. Not wanting to intrude, Goodall quietly left the remaining chimps and made her way back to camp.

A STRANGER COMES TO CAMP

During the summer of 1962 a stranger arrived in Gombe driving a Land Rover filled with camera equipment. His name, "Hugo Van Lawick," was emblazoned on the side of the vehicle. Van Lawick was a Dutch photographer with whom the National Geographic Society had contracted to document Goodall's work. Leakey had initially recommended him. The Society agreed and underwrote his travel expenses. Van Lawick was to provide still photographs and moving pictures of Goodall going about her daily routine. The Society hoped to produce a documentary film about Goodall, as well as to maintain a record of her ongoing activities.

Goodall was hesitant about the project despite Leakey's reassurances. Fearful that the presence of a photographer would frighten the chimps and destroy the progress she had made with them, Goodall was reluctant at first to permit Van Lawick to photograph her in the field. She initially agreed only to have pictures taken in camp. Yet, even she saw the importance of maintaining a record of her work. She hoped that David Greybeard would accept the new "white ape," and would calm the fears of the other chimps.

In her conversations with Van Lawick, however, Goodall soon discovered that they shared much in common. He had been born in Indonesia and educated in England and Holland. He had grown up with an intense love of animals that eventually led him to a career in wildlife photography. Like Goodall, Van Lawick had been intrigued by the possibility of working with African wildlife. He had first come to Africa in 1957 to work as a film assistant for Armand and Michaela Denis who produced a television program "On Safari." While working for the Denises, Van Lawick became friends with Louis Leakey and later made a film about the Leakeys' excavation at Olduvai Gorge. That experience led to Leakey's recommendation of Van Lawick to the National Geographic Society. It was clear to Leakey that Van Lawick was not only an excellent photographer, but that he possessed a real love for, and understanding of, animals. Leakey believed that those qualities would render him sympathetic to Goodall's work and enable him to accept the restrictions she was likely to impose on him. In a letter to Vanne, Leakey even suggested that Van Lawick would make a perfect husband for Jane.

To avoid any appearance of impropriety, the National Geographic Society assigned a chaperone to Goodall and Van Lawick. As a young unmarried man and woman, it would have been unseemly to allow Goodall and Van Lawick to be alone with each other in the wilds of Africa. Vanne agreed to serve in that capacity, though she did not arrive in Gombe until several weeks after Van Lawick had come.

DOWN TO BUSINESS

Van Lawick's first day at Gombe was memorable. Early in the morning David Greybeard arrived in camp. So as not to upset David by his sudden appearance, Goodall asked Van Lawick to stay in his tent. As Van Lawick watched through the tent flaps, David began eating bananas. When he had finished, he walked over to Van Lawick's tent, pulled the flap open, and stared at the startled photographer. He grunted and leisurely walked away. David Greybeard did not seem at all fazed by the presence of a stranger. Goodall later described his attitude, writing that it was as if David regarded Van Lawick as simply part of the furniture of the campsite and nothing to be feared.

Van Lawick's second day proved equally unforgettable. He and Jane witnessed the chimps hunt, kill, and eat their prey. That afternoon, as Goodall and Van Lawick were heading toward the Peak to watch four red colobus monkeys that had been separated from their group, Goodall noticed a male adolescent chimp climbing quietly up a tree next to the monkeys. The chimp moved cautiously along a branch. Three of the monkeys jumped away, leaving the fourth one alone. Seconds later, a small group of chimps charged from the bush and ran up the tree to join the first chimp. As they screamed and barked, they tore the lone colobus monkey to pieces in a matter of a minute. Too far away to film the kill, Van Lawick had to settle for shots of the chimps eating.

This piece of extraordinary good luck was to be the last Van Lawick enjoyed for a long time. After shooting for several weeks, he did not have nearly enough footage to complete the documentary. He had some astonishing footage, and had snapped some striking still photos of William, Goliath, and David Greybeard in camp. But whenever the chimps encountered Van Lawick in the wild, they fled from him as they had once fled Jane. Goodall tried everything she could think of to conceal Van Lawick from the chimps' view, but they did not fall victim to these deceptions. They recognized the sound of a camera whirling and clicking, and scattered whenever they heard it.

Van Lawick refused to give up. He trudged through the valleys, and sat for hours in the hot sun heat perched on the rocky slopes. Some days he saw nothing; on others, when he did catch glimpses of the chimpanzees, it was usually as they ran away from him. But after about six weeks Goodall noticed that chimps were becoming used to Van Lawick's presence. As Goodall had hoped, David Greybeard helped the process along. Whenever he encountered Goodall and Van Lawick outside of the camp, he came to see if they had any bananas. The other chimps watched, and a few began cautiously to approach Goodall and Van Lawick as David had done. Slowly, the chimps were coming to accept Van Lawick.

HUGO AND THE CHIMPS

Crouching to watch a group of chimps feeding in a nearby tree, Van Lawick had just begun to film when he felt his camera being pulled away from him by a black, hairy hand. Goodall had earlier discovered that David, Goliath, and William loved to chew on cloth and cardboard, and were particularly fond of sweaty clothing. David Greybeard had followed Van Lawick because he wanted the old shirt that Van Lawick used to cover his camera lens. Lawick engaged the chimp in a tug of war until the shirt ripped in two and David walked away with his prize. Van Lawick saw that the other chimps had stopped eating and watched this episode with great interest. Afterward, they became more tolerant of him and no longer fled the whirring and clicking sounds of the camera.

The presence of baboons, however, presented an ongoing problem that Goodall and Van Lawick had to confront. When Jane fed bananas to David Greybeard, Goliath, and William, several baboons now also insisted that she provided bananas for them. Goodall repeatedly tried and failed to discourage them. One day, as William, David, and Goliath were feeding off a large pile of bananas, a male baboon charged. William and David skittered away, but then David returned. Goliath continued to eat and ignored the commotion. David stood next to Goliath and began screaming and shouting at the baboon. The baboon lurched forward. David embraced Goliath, who by now had moved toward the baboon and begun to scream and wave his arms. The baboon made a slight retreat, but then lunged at David. This scene repeated itself. Each time Goliath moved toward the baboon, the baboon retreated but hit David. Van Lawick had captured the entire incident on film. Today experts consider this footage among the best records of an aggressive encounter between chimpanzees and baboons.

FAREWELLS AND RECOGNITIONS

The last 10 days of Van Lawick's visit provided him with the additional footage he needed to complete the documentary. He filmed chimps making and using tools and fishing for termites. At the same time, he continued to take still photographs of William, David Greybeard, and Goliath as they went about their daily activities. By the end of November, as he made preparations to leave Gombe, Van Lawick was so convinced of the quality and value of his material that he planned to ask the National Geographic

Society to fund a second trip the following year so that he could continue compiling the visual documentation of Goodall's work.

With his departure, Goodall was alone once more. Although not lonely, she had to admit to herself that she was not as contented to be alone as before Van Lawick had arrived. She began to realize how deeply Van Lawick appreciated and understood what she was trying to do. Van Lawick was a kindred spirit, and Jane had to acknowledge that she missed him. To ease her loneliness, Goodall buried herself in work. Her emotional bond with the chimpanzees, and especially David Greybeard, William, and Goliath, had grown stronger. More than ever, she felt that Gombe was where she was supposed to be and working with chimpanzees was what she was called to do.

A GIFT AND A LOSS

Christmas Day, 1962, was a day to remember. The morning began with a visit from William and Goliath, who howled with delight at the huge pile of bananas awaiting them under a small tree that Goodall had decorated with silver paper and cotton. Flinging their arms around one another, they patted each other on the back and danced around. Then the feast began, as Goliath and William, uttering cries of delight, settled down to the business at hand.

David Greybeard arrived later. Goodall sat next to him as he ate, and then slowly moved her hand to touch his shoulder, making a grooming movement. David brushed her away but not in anger, so Goodall tried again. This time, David allowed her to groom him for a minute or two. Then he gently pushed her hand away. Goodall understood and stopped, but she had established physical contact with a wild chimp. It was a Christmas gift to treasure.

After Christmas, Goodall prepared to return to Cambridge for another semester. The two weeks before her departure were sad. William became ill. His nose ran. He coughed terribly. His eyes watered. Once when Goodall had gone to check on him, she found him asleep near the camp. All that afternoon, Goodall remained nearby. She noticed that William had begun to urinate on himself, a behavior so unusual in chimps that she knew it meant his illness was worsening. She decided to spend the night. William ate sporadically and dozed off and on, all the time his body wracked by coughing and wheezing. Later that night it began raining, and Goodall watched as William huddled against the rain to keep warm, coughing now almost incessantly. Then, as the rain fell more heavily, William stopped coughing. A deathly silence followed.

Goodall stayed with William almost constantly throughout the following week. He remained close to the camp, making several nests for himself. Sometimes David or Goliath came to visit him, but William himself never ventured far from the camp.

Two days before Goodall was to leave for England, William stole a blanket from Dominic's bed and chewed it. Soon David Greybeard joined William in this activity. Then William placed the blanket over his head and made groping movements toward David. David stared at William for a moment, and then patted his hand. They left together, leaving the chewed blanket on the ground. Goodall hoped that this meant William had recovered from his illness. She never knew, for she did not see him again.

1963–1966

The years between 1963 and 1966 were productive for Goodall. She was beginning to make a reputation for herself in the scientific community. The National Geographic Society publicized her work with the chimpanzees, and Goodall and her chimps increasingly became the subject of television, film, newspaper, and magazine stories, including several accounts that Goodall herself had written.

In January 1963, Goodall returned to England to continue her doctoral studies. She knew the trouble to which Leakey had gone to get her into Cambridge. He still believed that earning a Ph.D. would make Goodall and her work more credible. It was important, he explained, that she be accepted into the scholarly community. If she did not pursue the degree, he feared that she would continue to face criticism and her work would be dismissed as insignificant, making it more difficult to secure funding.

By the spring of 1963, Goodall was back in Gombe. Hugo Van Lawick had also returned to shoot more footage and photographs for National Geographic. By now, too, more chimps visited the camp, among them the old matriarch, Flo, her daughter Fifi, who was almost three and half years old, and her son Figan, whom Goodall estimated to be about seven. The three were initially cautious and hesitant, but gradually they relaxed and became members of the Banana Club. Goliath and David still came regularly along with another male chimp, Evered, the son of Olly, whom Goodall believed was William's sister. Even crusty old Mr. McGregor came to visit and eat now and then. The chimps were becoming tamer around Goodall and Van Lawick, both of whom found them easier to approach.

Yet, the very success of the Banana Club caused some serious problems to arise. Goodall witnessed more instances of violence between the chimps and baboons as well as arguments and squabbles between the chimps over bananas. The dominant males took the best bananas for themselves, leaving the remains for others to fight over. Chimps trying to protect their haul of bananas also menaced Van Lawick and Goodall. In one instance, a chimp who thought Van Lawick had come too close to him threw a rock that almost struck Van Lawick in the head. Goodall realized that even the weakest and most passive chimp could injure or kill anyone in the camp.

Her solution to the potential danger was to have constructed a large steel cage that offered protection in the event the chimps got out of control. Goodall and Van Lawick also worked on an alternate method for giving bananas to the chimps through the construction of a feed box that would control the number of bananas to which the chimps could gain access.

FLO

In July 1963, Goodall and Van Lawick came upon an unusual sight: Flo was rushing to get to the bananas just as David Greybeard and Goliath entered the camp. Flo's behavior was unusual because she tended to stay clear of the males, especially Goliath. But on closer inspection Goodall noticed that Flo's rear end had turned a bright pink, a sign that she was ready to mate.

Goodall had already learned a great deal about the patterns of mating and childbearing among female chimpanzees. Females, Goodall knew, had only one child every five or six years. The children, even as they got older, spent a great deal of time with their mothers. Several other male chimpanzees had followed Flo into the camp that morning, before Goliath and David Greybeard arrived. Goodall had never seen many of these chimps. Although they were interested and aroused, none approached Flo. When David and Goliath arrived, they gathered some bananas. Then Goliath made a sweeping gesture with his hand and Flo presented herself to him. They mated. When Goliath had finished, Flo invited David Greybeard to approach and she also mated with him.

The next day Flo returned to the camp. More male chimps followed her. Flo proceeded to mate with every male who had come to the camp. This process continued over the next three weeks. Meanwhile, Flo's daughter Fifi tried unsuccessfully to protect her mother from the males. When she failed, Goodall noted, Fifi withdrew from the group, sullen, frightened, and unsure.

When her mating cycle ended, Flo was exhausted. She was covered with scratches, cuts, scrapes, and bite marks; two pieces of her ears were missing. The next time she came to camp only Fifi accompanied her. None of the male chimps paid her the slightest bit of attention. A few days later, as Flo was making her way to camp, several males frightened her and she climbed the nearest tree. David Greybeard went up to see her and everything was quiet. When David climbed down, Flo followed. Once on the ground she presented herself again, but the males showed no interest. Goodall always wondered what Flo was feeling at that moment.

During her mating cycle, Goodall noticed, Flo had one serious suitor: Rodolf, among the highest-ranking male chimps in the group. Rodolf walked beside her, and when she became tired or frightened, he laid a reassuring hand on her shoulder or put his arm around her. At night, Rodolf slept beside Flo, but never interfered with the attempts of other males to mate with her. After the mating cycle ended, Rodolf continued to care for Flo during the next several weeks, grooming her and traveling with her and her children.

One other aspect of chimp behavior that Goodall noted was that between an adult male and his mother. As she watched the young male chimps growing up, their relationships began to change. The males began to dominate the females. Then, while initially fearful of the older males, younger males begin to challenge their low ranking. As a chimp's testosterone levels rise, he becomes bigger, stronger, and more aggressive. But in the midst of all these changes, the male chimp's relationship with his mother never changes. Goodall noticed that many adult males were respectful and attentive to their mothers. And in cases of injury, the male often sought out his mother for comfort.

In August, Goodall published her first article, "My Life Among the Chimpanzees," in *National Geographic*. It featured numerous photographs of Goodall, Hassan, and Dominic at work, of the chimps making tools and carrying meat, and of David Greybeard and Goliath. The issue sold more than three million copies and introduced Goodall to the world. The essay marked the beginning of a long and successful writing career for Goodall and a longstanding collaboration with the National Geographic Society.

LOVE, MARRIAGE, AND A BABY

The episode with Flo pointed out the problems with the Banana Club. Goodall and Van Lawick wondered if a permanent feeding station was not now in order. With Hassan's help, they built a number of concrete boxes with steel lids that opened outward. They then sunk these boxes into the ground, the lids shut tight by wires. When the pins holding the handles in place were released, the wires became slack and the lids fell open. This system, Goodall hoped, would enable her to observe a number of different chimps who did not regularly come to the camp.

In December 1963, Kris Pirozynski, a Polish mycologist, joined Goodall at Gombe. Since Goodall was due back in Cambridge and Van Lawick had contracted to photograph a safari, Pirozynski offered to look after the camp and the chimps for the next four months. Hassan and Dominic also stayed on; both were excited at the prospect of studying chimps on their own.

By this time, too, the nature of Goodall's and Van Lawick's relationship had changed. They had fallen in love. Before committing to marriage, though, they decided it would be a good idea to spend time apart. They planned to meet again in Washington, D.C., early in 1964 when Van Lawick showed the film he had made about Goodall to members of the National Geographic Society.

As it turned out, Van Lawick could not wait. On the day after Christmas, 1963, Goodall received a telegram at her family home in Bournemouth. It was from Van Lawick, asking her to marry him. She immediately accepted and the two made plans to wed after Goodall was finished with her term at Cambridge and the Van Lawick film about her premiered in Washington. The wedding, which took place on March 28, 1964, was as joyous as it was unusual. The wedding cake was topped with a clay figure of David Greybeard, while color portraits of Flo, Fifi, Goliath, and the other chimps gazed down upon the guests.

The newlyweds cut their honeymoon short by three days to return to Gombe. Battling their way through flooded rivers, taking detours, and finally shipping the Land Rover by train, Goodall and Van Lawick at last arrived Gombe where they learned that seven weeks earlier Flo had given birth to a son. They christened him Flint. Flo cradled the child in her arms and carefully groomed his tiny body. Her other children, Fifi and Figan, always remained close by. Goodall took the birth to be a good omen.

MORE CHANGES

As Goodall and Van Lawick settled into their routine as husband and wife, Kris and Dominic reported some exciting news. A number of new chimps, including several females, now regularly visited the camp. Goliath also appeared to be losing his dominant status to Mike. This development was of particular interest, for Mike had previously been among the low-ranking males in the group. But in the relatively brief time Goodall had been away, Mike had asserted himself, largely by banging on empty kerosene containers and then charging at other chimps. Although this turn of events fascinated Goodall, she soon became apprehensive, realizing that Mike posed a danger to the entire camp. Instead of merely banging on the kerosene containers, Mike had begun to hurl them as he charged. One of the containers had struck Goodall, hitting her on the back of the head. Another narrowly missed Van Lawick's camera. Mike's use of the kerosene cans as weapons indicated to Goodall that he was of above average intelligence. Yet, something had to be done before he injured someone. To solve the problem, Goodall and Van Lawick began to bury the cans, leaving Mike with only stones and branches to use as weapons.

Goliath was not about to relinquish his high-ranking position easily. Throughout the next year, he and Mike continued to have confrontations. Finally, one day, the two engaged in a complex and spectacular display to determine which was superior. Goliath lost and submitted to Mike's supremacy. He sat down next to Mike and began to groom him. After a few minutes, Mike did the same to Goliath. This ritual continued for more than an hour. From then on, Goliath accepted Mike's superiority, and, in time, the two became very close companions, often eating or resting together. For Goodall, the experience showed how quickly chimpanzees put aside their rivalries and reestablished lasting bonds of mutual respect and even love.

At about the same time that Mike and Goliath worked out their differences, other problems arose at the camp. As the chimps became less fearful of people, they began to run amuck. Kris told Goodall that some chimps had taken to stealing his clothing and bedding, while others destroyed chairs and tent flaps by chewing them to pieces. One chimp had even figured out how to pry open the steel lid of the box that held the bananas and help himself to the contents. Most astonishing of all, some of the bolder chimps raided the huts of the nearby fishing villages to steal clothing.

Fearful that chimps might be hurt or killed, or that they might hurt or kill a person, Goodall acknowledged that the Banana Club may have come to an end. She, Van Lawick, and Kris debated what to do. They decided to move the camp deeper into the valley. Goodall and Van Lawick attracted the chimpanzees to the new site by waving an empty banana box at David Greybeard, who then acquainted the group with the new campsite.

The move was a success. With the camp now situated further into the valley and away from local settlements, more chimps visited more often. Goodall happily noted that much of the fear and apprehension the chimps had earlier displayed now disappeared. In general, the chimps seemed more at ease.

THE GOMBE STREAM RESEARCH CENTER

When Goodall first came to Gombe in 1960, she never dreamed that within four years she would have students and researchers from around the world to help her with her field studies. But by April 1964, Edna Koning had arrived and helped transcribe Goodall's notes. Sonia Ivey came in December to study baboons, but also worked as Goodall's secretary and assistant. Her staff of Africans continued to grow and managed the day-to-day operations of the camp. Visitors included staff from the National Geographic Society, the governor general of Kenya, and Goodall's mother, Vanne. Two distinguished scientists, Hans Hass, a pioneer in the study of biology, and Irenäeus Eibl-Eibesfeldt, a noted ethologist, also paid a visit to Gombe. Encouraged by the positive reports of Hass and Eibl-Eibesfeldt, Goodall applied to the National Geographic Society for additional funding. The Society agreed.

Even with the additional help, workdays at the camp frequently lasted well into the night. The group became so immersed in the lives and habits of the chimps that they talked of little else. And there was plenty to talk about. In addition to Flo, two females had given birth. Melissa bore her son, Goblin, in September, and in November, Mandy bore her daughter Jane. Almost 45 chimps now visited the camp on a regular basis. As they became more accustomed to Goodall, they allowed her to follow them deeper into the forest where she could observe them for longer periods of time.

The group continued to improvise solutions to the problems that continually arose. By now, everyone kept their belongings in a wood box. To protect the tents, Hassan fashioned some tent poles from sturdy tree trunks that made the tents more difficult to tear down. Yet, however vigilant they were, Goodall and her team simply could not fill the banana boxes fast enough. Hassan made more boxes, but in the end they only replaced those that the chimps had destroyed. Male chimps also accosted those carrying bananas to the boxes and hijacked the fruit.

Amid the almost constant activity at the camp, Goodall and Van Lawick discussed the creation of a permanent research center. In March 1964, Goodall had written to the National Geographic Society to propose the establishment of a center. Among her supporters were Louis Leakey, Dr. Hass, and Dr. Eibl-Eibesfeldt. By November, the Society had agreed to fund the purchase of a site and the installation of a few prefabricated buildings in the area Jane had selected. The site provided a magnificent view of the mountains and the lake. By December, a concrete foundation for the largest building had been poured and construction begun. Once up, the buildings were covered with grass and bamboo. By early 1965, the Gombe Stream Research Facility was operational. It consisted of a large working area, sleeping quarters for Edna and Sonya, a kitchen, and small storage area. Goodall and Van Lawick lived in a smaller building, while another small structure served as a storehouse for bananas. Within three days of its completion, the Center was overrun with chimpanzees.

A few weeks later, Dr. Leonard Carmichael, Dr. Melvin M. Payne, and Dr. T. Dale Stewart visited to evaluate the center on behalf of the National Geographic Society. Goodall thought that even the chimps sensed the importance of their visit. As Goodall began to outline her progress and her plans for the future, Payne unexpectedly exploded in anger at what he believed was her insolence. She also had heated arguments with Carmichael and Stewart. But in the end, they came to appreciate her determination and maturity and agreed that somehow the Gombe research center needed to become a permanent facility.

For the time being, though, Goodall was thrilled with the new center. It was far more comfortable than the camp had been, and offered greater security. So it was with great regret that she and Van Lawick made plans to leave: Goodall to return to Cambridge to complete her doctoral dissertation and Van Lawick to start another photography assignment.

After they left Gombe, both Goodall and Van Lawick realized the terrible mistake they had made by encouraging some of the young chimps to touch them. Goodall had even played with Flint. The center would draw more students. A chimp that had grown comfortable around humans might also sense weakness and fear, creating a potentially dangerous situation. Goodall and Van Lawick decided to establish a rule that under no circumstances were inexperienced students to have physical contact with any of the chimps, no matter how gentle they might appear.

Goodall was pleased with and proud of the Gombe Stream Research Center. But success came at a price. Publicizing and raising additional funding for the center meant spending more time away from Gombe. She regretted her frequent absences, but also recognized their necessity if the center were to continue the work she had begun. So it was with a heavy heart that she left Gombe at the end of 1965. She would not return for an entire year.

FINISHED STUDIES

Despite Leakey's insistence that she earn a doctorate, Goodall had always hated the idea, resenting the time away from Gombe and the chimps. She once confessed to an interviewer that she never wanted the Ph.D. Yet, she dutifully continued her studies and received her Ph.D. in ethology, the comparative study of animal behavior, in 1965, becoming only one of eight persons to receive a doctorate from Cambridge without first earning a bachelor's degree.

Goodall had to admit that she was proud of her accomplishment, but her time in Cambridge had not been easy. Many of her fellow students thought her a fraud whose single claim to fame was an article published in *National Geographic*. Because Goodall's picture had been on the cover of the issue in which her essay appeared, others viewed her as a "pin-up" girl for the Society. Many of Goodall's colleagues also thought her research was superficial, charging that she merely related anecdotes and did not do scientific research. Her dissertation adviser, Robert Hinde, was among those who questioned the professionalism of her methods. He was especially critical of Goodall for naming her subjects instead of assigning them numbers. She may also have been guilty of one of the most serious blunders a scientific researcher could commit: anthropomorphizing animals by imagining them to have human characteristics that they could never possess. Goodall remembered that Hinde: "would suggest things like, 'When you get back to Gombe'—and this was after I had spent only eighteen months there—'you should measure the distances between where the chimps are feeding and the level they are in the canopy.' . . . I was lucky that I was never going into these things for science. And I didn't care about the PhD, it didn't matter. I would listen, I just wouldn't do what they said."[2] Despite his objections, Hinde in the end approved her thesis, but not without a struggle. When he returned her dissertation to her after his first reading, Goodall found that Hinde had crossed out the genders of the chimps, replacing the pronouns "he" and "she" with "it." Furious, Goodall restored all the gendered pronouns and kept the chimps' names. Hinde relented, and the thesis stayed as Goodall wanted it. Recalling that episode years later, Goodall explained: "How they would even want to deprive them of their gender I can't imagine. But that is what it was, animals were 'it.' Makes it a lot easier to torture them if they are an 'it.' Sometimes I wonder if the Nazis during the Holocaust referred to their prisoners as 'its.'"[3]

A GROWING REPUTATION

In 1965, Goodall turned 31 years old. She had already accomplished a great deal more than she had ever imagined. Between March 1965 and May 1966, Goodall fielded a flood of offers as her reputation continued to grow. Besides writing for *National Geographic*, she had submitted to Houghton Mifflin a book-length manuscript that detailed her early work with chimpanzees. She was also negotiating with Houghton Mifflin to write a second book.

In December 1965, Van Lawick's film, "Miss Goodall and the Wild Chimpanzees," debuted on American television, documenting her groundbreaking discoveries and greatly enhancing her reputation. She was, according to Dr. Leonard Carmichael, chairman of the National Geographic Society Committee for Research and Exploration, "the most qualified person in the world today to speak on the subject of chimpanzee behavior."[4] Carmichael's ringing endorsement of Goodall and her work helped to silence some of her critics.

In February 1966, Goodall undertook a successful lecture tour in the United States, speaking on three occasions to sold-out audiences in Washington, D.C. In March, she went to Bournemouth to give a lecture, before returning to the United States. The lecture tour concluded in April 1966, with Goodall's visit to the San Diego Zoo, which was celebrating its 50th anniversary. Among her duties was cutting a giant 320-foot-long birthday cake. Later that month, formal notification arrived from Cambridge that she was now known as Dr. Jane Goodall.

As pleased as she was that the world had begun to take respectful notice of her work, Goodall yearned more than ever to return to the comparative isolation of Gombe. She and Van Lawick finally did return in the early summer of 1966. Upon their arrival, they learned of new successes and serious problems.

The most critical problem that greeted them was the shock at how much the chimps had changed in a year. Incidents of intimidation and fighting had increased among the chimps, driving many to seek permanent refuge in the camp. Goodall decided that the way to persuade them to leave was to limit the supply of bananas. Hassan constructed new banana boxes out of high-tech materials sent from Nairobi. These boxes could only be locked and unlocked electronically, thereby preventing the chimps from breaking into them.

Goodall also was disturbed and alarmed to find out that some of the students had disobeyed her rule not to permit the chimps to touch them. Goodall repeated her warnings about the potential danger of such acts and requested that they stop immediately.

Hoping that the problems were solved, even if only temporarily, Goodall and Van Lawick prepared to depart for the Serengeti in July 1966, he for a photographic assignment, she to research primates in the region. As she drove away from the Gombe Stream Research Center, Goodall could scarcely have realized the importance this trip was to have in shaping her already brilliant career.

NOTES

1. Wolfgang Kohler, *The Mentality of Apes* (London: Routledge & Paul, 1925), p. 9.

2. Scientific American, "Profile: Jane Goodall," http://www.sciam.com/1097issue/1097profile.html, December 7, 2001 (accessed July 1, 2004).

3. Scientific American, "Profile: Jane Goodall," http://www.sciam.com/1097issue/1097profile.html, December 7, 2001 (accessed July 1, 2004).

4. Jane Goodall, *Africa in My Blood: An Autobiography in Letters, The Early Years*, ed. Dale Peterson (New York: Houghton Mifflin, 2000), p. 334.

Chapter 8

EXPANDING HORIZONS

During Goodall's year-long absence, the center remained open and in the capable hands of Edna and Sonia. The two had, by now, learned how to take accurate field notes of their own and helped to maintain Goodall's work. As news of the Gombe Stream Research Center spread, Goodall began receiving applications from students all over the world to become research assistants. Soon the number of people working at the center expanded, as did the focus; in addition to studying the chimpanzees, some assistants were branching out on their own to study baboon and red colobus behavior. By 1972, more than a hundred people were helping with the study of 54 chimpanzees in addition to the baboons and monkeys.

All pitched in willingly with the work, often toiling days and nights on research. The majority of students Goodall recruited were from Stanford where she was teaching and Cambridge where she attended school. Many of the students, unlike Goodall when she first started, came armed with bachelor's degrees and plans to stay for a year. At the year's end, several requested to stay on for an additional year in order to work on more specialized areas of chimpanzee behavior.

A GUIDING SPIRIT GONE

But one person was missing throughout the building of the center: Louis Leakey. At first, he chose not to visit Goodall because he did not want to interfere with her work. Over time, his declining health made it difficult for him to travel to Gombe. However, the two kept in constant contact, writing each other as often as possible. Leakey also kept in touch with Vanne. He made a point of stopping at the Birches between trips to Kenya and the United States. While at Bournemouth, he and Vanne would go to plays and the ballet. He often attended church with Goodall's sister, Judy. In the late 1960s, Vanne began helping Leakey with his writing, editing his many books and articles. She also coauthored a book with Leakey, the 1969 work, *Unveiling Man's Origins*.

Leakey's confidence in Goodall's work never wavered. He was tremendously proud of her accomplishments, much like a father with his daughter. One day while visiting Vanne, Leakey predicted that someday a statue would be raised to honor Goodall. Another guest exclaimed, "Oh no—to you first, Dr. Leakey." Leakey thought a moment then replied, "No—to Jane."[1] In the autumn of 1972, the letters between Goodall and Leakey came to an end. Leakey had died of a heart attack at a London hospital with Vanne at his side. Goodall's mentor, the man who offered her the keys to her dream, was now gone forever. For Goodall, the best way to honor his memory was to dig deeper and harder to prove Leakey was right after all.

LIFE AT GOMBE

Over time, the research center also expanded its facilities; eight additional sleeping houses were built at the observation point in the valley, known as the "Upstairs." Here, placed out of sight, observers could quietly watch and record the chimps. Three larger buildings were built near the beach, with three more additional sleeping quarters for those students studying the baboons and other monkeys. The African staff, which was increasing every year, had their own quarters that were so large it constituted a small village. This area was soon nicknamed the "Downstairs." The facilities were basic; but for Goodall, the researchers, and the staff, they were more than adequate.

The surroundings may have been simple, but some amenities prevailed. The student housing consisted of little more than prefab metal cabins, the windows protected by screens to prevent chimps from crawling in. Bathing facilities were provided by the lake. A dining hall was built near the beach. Every night the students would dress for dinner and wait until everyone was seated. They then dined on rice, cabbage, or fish. If the cook staff had gone to Kigoma, there might also be fresh fruit and meat. After each meal, students might present an informal talk on their research. Other times, if the electric generator was working, they might listen to music on a cassette player. Sometimes Goodall arranged a special event or party; beginning in 1972, a weekly movie was instituted. It was

Goodall's hope that a "spirit of Gombe," a kind of cultural ideal would spring from the exchanges of knowledge and observation. By now, the study of the chimps had evolved into a community effort and resource. Not only were students learning, they now worked together as a unit for a greater purpose.

By the early 1970s, the majority of chimps were visiting the feeding station, which now was housed in a building of its own. Inside were bins that could be electronically opened and closed depending on the circumstances. Some researchers monitored attendance rates at the station every day, working on charts that showed group structure and activities, vocalizations, and gestures. More emphasis was put on tracking individual chimps and groups in their own habitats away from the station. As Goodall later recalled, there was so much information to be learned, there was no way one person could have adequately done the job.

Like Goodall, many of the students had chimps that they were particularly drawn to. Typically male chimps sought out male students, while female chimps were more comfortable with female researchers. For many students there formed a liking and respect for the chimps. Even years later, former students explain that the nature of their relationship with the chimps was difficult to describe. All agree that the nature of their dealings with the chimps was not forged on interaction or favors, but a kind of mutual attraction. All were certainly aware of Goodall's own relationships with David Greybeard, Flo, William, Goliath, and many others. None of them believed that they would ever experience such a relationship. When it happened, many began to understand Goodall's contribution in building a bridge between humans and animals that allowed others to walk across too.

CHANGES

The year 1967 was a year of many changes for Goodall and the Gombe Stream Research Center. Probably the most important was the Tanzanian government's decision to put the area under the watchful eyes of the Tanzanian National Parks. This would not only ensure the protection of the area, now known as the Gombe National Park, it would offer Goodall increased resources to work with. The Game Department scouts who had once worked with Goodall when she first came to Gombe were now replaced by National Parks Rangers. Housing was built for the Rangers in the southern section of the park. Goodall, working in conjunction with the National Park officials, began collaborating on a number of projects including the building of a second feeding station in the southern area for tourists and other visitors. Student researchers were already at work there trying to habituate the chimps as Goodall had over 25 years ago.

That same year Goodall also learned that she had been appointed Scientific Director of the Gombe Stream Research Center. But her greatest joy came with the birth of her first and only child, Hugo Eric Louis Van Lawick. The child bore an impressive pedigree; he was named not only after his father, but Goodall's Uncle Eric and her mentor, Louis Leakey. By the age of three, the child, then known as Little Hugo, had acquired a nickname he is still known affectionately by today: Grub. Grub's first months were divided between the Serengeti where his father was on assignment and Gombe. After that, the family settled in at the Research Center and Goodall's life took a dramatic turn both personally and professionally.

Goodall's heightened perspective in observing Flo and other female chimps interact with their youngsters influenced her own ideas about child raising. During her pregnancy, Goodall and her husband vowed to raise their child in a manner similar to Flo. Flo out of all the females was admired by Goodall, who thought her a model mother. She never left her children behind. She rushed to help them if they cried or got hurt. Her love for her children was demonstrated repeatedly through constant hugging, caressing, and grooming them. Even after the children got into mischief, Flo might discipline them, but then always let them know she still loved them. Goodall was especially drawn to Flo's willingness to play with her children. Flo was reassuring, patient, and loving, qualities that Goodall hoped to have in raising her son.

AN UNUSUAL UPBRINGING

Pushing aside conventional child-rearing practices, Goodall used a variety of techniques in raising Grub. Her influences included Vanne, Flo and her children, and the popular Dr. Benjamin Spock, whose best-selling book on childcare was considered the "bible" of how to raise children. At a time when most women bottle-fed their infants, Goodall instead breast-fed her son on demand for a year. Upon hearing Grub's cries, both Goodall and Van Lawick rushed to their son's side, hugging him and reassuring him. Van Lawick at one point photographed a sequence showing Goodall with Grub as he learned to walk. In a photo album they sent to their parents, Van Lawick also included a sequence of shots of Flo and her new infant Flint taking his first steps. Remarkably, the sequences varied little from each other. At one point, Goodall and her husband, fearful that the chimps might attack the infant, built a special cage painted bright blue, with colorful birds and stars hung from the cage's ceiling. Grub was placed in the cage where he could see the comings and goings around him and still be protected from danger.

Now raising a son, Goodall's daily routine centered more on her child than on her work. Goodall realized with some sadness, the peaceful isolation and excitement of her first years at Gombe were gone forever. But she knew those feelings were selfish. She recognized that with the creation of the center and the many people now working with her, her research had taken great strides in ways that she could not have done alone. For now, her mornings were spent at the house on the lakeshore where she wrote up her research findings, reports, and proposals, as well as tended to the ongoing administrative work that needed to be done. Grub was usually down at the beach, in the company of a staff member. At some point during the day, Goodall took an hour to visit the feeding station watching for chimpanzees.

One thing Goodall learned by watching Flo and her offspring was that raising a child should be fun, not work. So every afternoon, Goodall stopped work for the day to spend time with Grub. The two went exploring, swimming, or just sat together and played. Grub often ran naked about Gombe; the weather was so warm that he needed few clothes. For Goodall, being able to touch and hug her son gave her a sense of her child's energy and the purpose of her own life. Taking a cue from Flo, Goodall would distract Grub when he was about to do something dangerous or mischievous; as he grew older, Goodall explained the consequences of any wrongdoing and held her son close.

Both Goodall and Van Lawick noted that Grub and Flo's son Flint were progressing at about the same pace. Goodall later joked that her son was merely trying to keep up with Flint and the other chimp youngsters. At the same time, she began to notice that Flint was becoming Grub's adversary. The chimp would jump on Grub or pull his hair. Goodall admits that at the time she felt an irrational anger toward Flint, and then realized how Flo must have felt when breaking up altercations between Flint and other chimp youngsters. However, Grub did not like the chimpanzees. When asked later if his dislike was out of jealousy, Grub replied no, that his aversion to the chimps was more out of fear than anything else.

Grub's parents also saw to it that he was comfortable in both the European and African worlds. Besides learning English, Grub also learned Swahili and KiHa, the most prominent languages in Gombe. His playmates came from the villages around the research center. Grub had a Christian upbringing, but was also comfortable with African spiritual beliefs such as witchcraft and rituals that helped protect children and their village. He had many happy hours playing with the African children.

People visiting Gombe were shocked and dismayed by Goodall's unconventional child-raising methods. She was criticized for bringing up a feral child. The building of the cage was another concern; according to the conventional wisdom, children were to be taught to become independent of their parents. Visitors to Gombe warned Goodall that not only would her child suffer deep psychological wounds from being shut in a cage, Grub would also become clingy and dependent. Grub's nakedness was also a concern. On a trip to Washington, D.C., Goodall, Van Lawick, and Grub were staying at a friend's house. One day, Grub, wearing no diapers, urinated on the couch as their host watched in horror. But, like her research, Goodall was committed to her style of rearing Grub, no matter what people thought.

While some people thought Goodall's raising of her son had required a deep sacrifice, Goodall saw the experience in other terms. While there was a sacrifice in giving up a previously hectic and long work schedule, Goodall saw her time with Grub as crucial. From her observation of the chimps over the last several years, Goodall understood how critical the early years of childhood were. Giving Grub as much of her time as possible was important not only in teaching and nurturing him, but in helping her witness the world through her child's eyes. Not to do so would stand as a hypocrisy of sorts. As Goodall later wrote, what was the point of watching and documenting chimp behavior, especially actions that might be of some benefit to humans? Why bother to study chimp behavior if, in turn, she ignored those lessons? More important, how could anyone take her own work seriously if she talked about the importance of the mother-child relationship, while ignoring her own child?

GOING AGAINST THE TIDE

By the late 1960s and early 1970s, Goodall was becoming a stronger presence within the scientific community. She was spending more time away from Gombe, giving lectures and talks. In 1970, she was appointed to the faculty of Stanford University as a visiting professor of Psychiatry and Human Behavior. But again a price had come with the added visibility. She rarely followed the chimps; students and field assistants now took over the observation and recording of field notes. Goodall's main duties consisted of running the center as its head administrator and scientist. Sadly, Goodall realized that almost every aspect of chimp behavior was now being studied by others rather than by her. Goodall knew that these changes were necessary for the growth of the center, but it also made her feel more like an intruder as she listened to the daily findings and talk of the other students and researchers.

During this period, another turbulent change was occurring in the West, as the growing women's liberation movement became a more visible presence in American and European life. Many women saw Goodall as one of the great pioneers in women's liberation. After all, she had broken into a previously male-dominated field and was successful at what she did. But Goodall was uncomfortable in that role. In a series of lectures given in the United States, she demonstrated that she had embarked on a completely different mission: to promote the traditional role of women in society. Using her studies of Flo and other female chimpanzees, Goodall argued that motherhood, not career, was the single most important force in human society.

To bolster her arguments, Goodall explained how good chimp mothers raised confident children. These chimps were socially adjusted, agile, and competent. As they grew older, they drew on their experiences as youngsters to move in and adapt to chimp society. Motherhood, then, was important not in terms of establishing a matriarchal society, but in helping mold the future. Goodall described how her observations of Flo and her children taught her "to honor the role of the mother in society."[2] Later in an interview with a *New York Times* reporter, Goodall flatly stated that "I never, ever, *ever* put my career before my child."[3] Her statement angered many who felt that Goodall was a traitor to the movement. However, Goodall also charged that governments needed to be more responsive to those mothers who had to work to support their families and could not afford expensive day care.

Many women were not particularly happy at being compared to chimpanzees, or at the thought that chimps should serve as role models for the modern twentieth-century woman. At a lecture at Yale University in 1972, Goodall was asked by a reporter if she believed that the submissive behavior of the female chimps was an adaptive behavior. Goodall said "yes" and described how a young male student had asked the very same question at Gombe. He had been studying the chimps and came to the same conclusion as Goodall. When he reported his conclusions to the group, they were very unpopular. Goodall's response to him and the reporter was the same: if researchers were looking for examples of women's liberation, then studying chimps was not the animal to research.

Goodall's continued criticisms of the women's movement made her increasingly unpopular with feminists. But for every feminist who ridiculed Goodall's beliefs and accused her of reinforcing the traditional role of women, there were numerous young mothers who thanked her for giving them the courage to spend time with their children. For Goodall, their responses made her lectures all the more worthwhile.

A TRAGIC TIME AT GOMBE

Since her first days in Gombe, Goodall had grown to love certain chimps. David Greybeard, Goliath, William, his sister Olly, Flo, her daughter, Fifi, and son, Flint, were for Goodall very special animals. There were others, too, that she had come to care for deeply. She realized that her feelings for these chimps were completely one-sided; the chimps did not respond to her in the same way. But each had in their own way helped her in her studies; each had reached out to communicate, if only briefly, their trust in Goodall. But all relationships come to an end sometime, and as the years passed, Goodall watched with great sadness the passing of her first companions.

In 1966, a polio epidemic swept through Gombe. Polio is a viral infection that is spread orally and can cause severe crippling and other handicaps. Because of the close relationship between humans and chimpanzees, chimps often are afflicted by the same types of disease. The epidemic at the reserve was in all likelihood given to the chimps by humans. The first sign came with the illness of Olly's young infant, Grosvenor, who suddenly lost control of its limbs. A day later, Goodall saw Olly carrying her dead child. Olly became so distraught over the infant's death that she carried the lifeless body around with her, either holding the infant or slinging him over her shoulder. Her young daughter, Gilka, even tried to groom and play with her dead baby brother. What Goodall and her husband did not realize was that Grosvenor was the first victim of the epidemic.

Once the cause of the sickness was discovered, Goodall radioed Louis Leakey who arranged a donation of the polio vaccine not only for the humans at the center, but for the chimpanzees as well. The vaccine would be given to the chimps in bananas. But for some chimps, help came too late. Mr. McGregor, one of the first chimps Goodall met, was among those who could not be helped.

In November, Goodall noticed some of the chimps staring at a low bush making soft sounds. When dusk fell Goodall and Van Lawick approached the bush fully expecting to see a dead animal. Instead they found Mr. McGregor trying to pull down berries with his arms. His legs were useless, as were his sphincter muscles. Covered in excrement and blood, swarms of flies danced around him. The chimp clearly was in grave distress. For the next two weeks, Goodall kept a careful watch over the chimp, hoping for some sign of recovery. They began taking him food, which they pushed up with a stick to Mr. McGregor's nest. Having let down his guard, Mr. McGregor let Goodall tend to him. She squeezed water from a sponge and had the chimp treated with insecticide to relieve the constant misery from the flies.

One evening, as they came to feed the old chimp, they found him lying on the ground. He had fallen out of his nest and dislocated his arm. Goodall realized that there was nothing more they could do for Mr. McGregor. They fixed a small nest for him, which he settled into and fell asleep. When Goodall checked on him later in the evening, Mr. McGregor opened one eye, gazed at her and then fell back asleep. The next morning Goodall offered a special treat to the ailing chimp: eggs. As Mr. McGregor was enjoying his meal, Goodall shot her old friend. She wept afterwards.

The effect of the epidemic on the chimp community was devastating. Several chimps were afflicted and six eventually died. Those who recovered were disabled for the rest of their lives. Some, in their weakened condition, were isolated from the community because they could not keep up. Some were killed by other, more aggressive chimps, while others simply disappeared into the forest. But their were some chimps, both family members and friends, who protected and helped the crippled chimps eat, make nests, and travel.

Many criticized Goodall for interfering with nature instead of letting nature take its own course. Even decades later, Goodall still defends her actions during the polio epidemic and other health crises at Gombe. For her, it was time to act, not to observe, choosing empathy over objectivity. Other scientists working with "primitive" peoples are often put in the same position. Given disease or disaster, one must make a choice: to help provide care and comfort or to stand by and watch, leaving the group free of Western influences and technology that might aid and comfort them.

This was not the last time Goodall would step in. When chimps were sick, she offered them antibiotics in bananas. If chimps came down fungal growths or other maladies, Goodall had biopsies done and dispensed medication to help fight the illness. When chimps lay dying, Goodall tried to make their last days easier by offering special foods, helping them make nests, or taking food to them as she did with Mr. McGregor.

Goodall recognized her actions would more often than not meet with disapproval from the majority of scientists and that nature should be allowed to run its course. She had already battled with many over her close contact with the chimps, especially the feeding by hand and the use of the feeding stations. Accused of interfering in a wild animal's life, that her actions colored her research, Goodall stood firm, arguing that humans have always interfered with nature, usually in a negative way. Instead, more positive influence was needed; humans needed to help rather than hurt. At many of her lectures, Goodall was repeatedly asked why she interfered. For her, it sometimes boiled down to a gender difference. Goodall believed that men tended to protect from interference the right to a life with self-fulfillment and dignity. Women, however, responded out of care, nurturing, and a desire to help alleviate—if only for a short time—the pain and suffering they encountered. This was never clearer than when Goodall lectured to a group of physical therapists, most of them women, about the polio epidemic. The first question asked of her was not why she decided to help, but did she try to help.

OLD FRIENDS LOST

Another tragedy visited Gombe at the beginning of 1968 when a flu-like epidemic swept through the park. Two mothers died; their deaths, in turn, resulted in their youngsters falling into a deep depression and later dying from malnutrition. In some cases, the chimps became so ill that Goodall and her staff had to put them down. Other strange deaths occurred: a chimp fell out of a tree and broke his neck. There was also a rash of poisonings at the southern sector of the park that killed a number of baboons.

But for Goodall, the most devastating loss came with the death of David Greybeard. Goodall was away from the center when she received news that David had gone missing, and had possibly come down with pneumonia. Many thought that eventually he would come back to the camp and kept a vigil, but as the days and weeks went by, their hope was replaced by sadness. By July 1, when Goodall came back to Gombe, there still was no sign of David. Goodall knew deep in her heart that David Greybeard, her first companion and her favorite, was dead. She was grief-stricken in a way she did not think possible.

Years later Goodall still speaks of David Greybeard. In one interview she stated that "my own relationship with David was unique—and never will be repeated.... When David disappeared ... I mourned for him as I have no other chimpanzee before or since."[4] Goodall realized that without David's help, she would have never penetrated the world of the chimpanzee. He was her window; he not only broke down barriers, but offered reassurance. Goodall and David also communicated on a level not found with the other chimps. And it was David who provided Goodall with her first important discoveries. To this day, Goodall continues to acknowledge David Greybeard's contributions to her work.

Just as tragic was the death of Flo in the summer of 1972. She continued to visit the camp with her children Fifi and Flint. Figan by this time had gone off with the older male chimpanzees. Flo had given birth to one more child, a girl named Flame. Sadly, after six weeks, Flame disappeared, perhaps as a result of illness. Flint by this time became increasingly dependent on his mother, as if reverting to infancy. He begged for his mother to take care of him. Even as she grew frailer and experienced difficulty walking, Flint still begged to be carried on her back. He always remained nearby, afraid to let his mother out of his sight. Flo by this time had become just as dependent on her son.

In August, one of Goodall's assistants came to her door with sad news: Flo had been found dead near the Kakombe Stream. She appeared to have died just before attempting to cross the rushing waters. Goodall stopped what she was doing and went immediately to the site. Upon seeing Flo's body, she gently turned her over and looked into another old friend's face, one that she had come to love dearly over the last 11 years. Flo lay there peacefully, without any sign of fear or pain. It was as if her heart finally had given out. In honor of her old friend, Goodall kept a vigil that night to prevent Flo's body from being torn apart by predators. She also wished to protect Flint, who remained nearby, from discovering his mother's body in shreds. Flint was clearly in deep shock and grief; Goodall recalled that he spent hours hunched near the stream bank where Flo died. From time to time he pulled at her hand, trying to pull her back to life.

As she sat in the moonlight, Goodall thought about Flo and her family. By her estimation, Flo had lived for almost 50 years roaming about the Gombe hills. Even if Goodall had not encountered her, she believed that Flo's life would still have been significant and with purpose. She had taught Goodall many valuable lessons, especially the importance of a mother and child relationship and the joy and contentment it offered.

To Goodall's dismay, Flint never recovered from his mother's death. Three days after Flo's death, Flint became increasingly lethargic, and then disappeared for six days. When he finally did come back to camp, his condition had seriously deteriorated. He was clearly suffering from depression and had taken to isolating himself from the rest of the chimps. Two weeks later, Flint died. He was eight-and-a-half years old. Goodall later wrote that his death had been caused by physical and mental distress due to his mother's death. But for all the scientific jargon, Goodall knew deep inside that the reality was much simpler. "Flint," she later said, "died of grief."[5]

Goodall was still grappling with her own grief over Flo's death a month later. To pay homage to Flo, Goodall wrote an obituary that was published in the *London Sunday Times*. The obituary was not only rare in that a major newspaper published it, but in its tone, which was at once poignant and celebratory:

> Flo has contributed much to science. She and her large family have provided a wealth of information about chimpanzee behaviors—infant development, family relationships, aggression, dominance, sex. . . . But this should not be the final word. It is true that her life was worthwhile because it enriched human understanding. But even if no one had studied the chimpanzees at Gombe, Flo's life, rich and full of vigor and love, would still have had a meaning and significance in the pattern of things.[6]

The deaths of Flo and Flint were among the sadder events in the history of Gombe. Flo's mothering techniques and social behavior had taught Goodall so much. And Flint, in his own way, also contributed to Gombe, becoming the first wild chimp infant whose development Goodall and her researchers were able to study in detail. With their deaths, and the death of David Greybeard, a remarkable chapter in the history of Gombe had ended.

The deaths of David, Flo, Flint, and the other chimps were heartbreaking; but during the same period, there was human tragedy as well. In 1968, one of the center's American students, Ruth Davies, was killed. She fell to her death after stepping into a hidden precipice in the valley. She had been following several of the male chimps, including Mike and Goliath. When her body was found, a small tape recorder lay next to her, which provided Goodall with an account of her last hours. Ruth had been traveling with Van Lawick in the southern sector of the park. Trailing the chimps had left her exhausted; on the tape one could hear her struggles for breath. Somehow she had slipped and fallen over a cliff face that had been hidden from view by heavy foliage. She lay undiscovered for five days.

Ruth Davis loved Gombe and the chimps; she had a passion for knowledge and was keenly interested in the behavior of the chimps. Because of her deep feelings for Gombe, her parents decided that Ruth should be buried in the Gombe hills. They later told Goodall that despite their grief, they were happy to have seen the land that meant so much to their daughter. Goodall understood, for Ruth had often expressed how some of her happiest days had been spent in the quiet and solitude of the forests and valleys. Goodall often visited her gravesite, feeling the quiet and gentle presence that Ruth, in life, imparted.

Ruth's untimely death also led to a change in procedures at Gombe. Goodall decided that under no circumstances would a student be allowed to tramp about without the company of a local Tanzanian. Even though Ruth had died instantly, Goodall and others at the center worried during their search that Ruth had been seriously injured and could not summon help. If she had been with a guide, at least the other person could have gotten aid much sooner.

Goodall then hired a number of men from the local villages. Among her first official field assistants were Hilali Matama, Eslom Mpongo, Hamisi Mkono, and Yahaya Alamasi, many of whom stayed on at Gombe for years. Goodall soon discovered that many of them had the capability to become first-rate field researchers and trained them how to conduct field studies. The men soon became an indispensable part of the Gombe research team.

A LOVE LOST

While Goodall could be proud of her mothering, her marriage to Van Lawick was suffering. The couple increasingly spent time apart, as Van Lawick continued to accept assignments that took him away from Gombe and Goodall began traveling more and more on long speaking engagements. There was also a basic incompatibility, but both believed that the other partner would change. Soon the two fell into a state of constant arguing and quarreling.

Clearly Goodall loved her husband, but as one friend later explained, the relationship was a high-pressure one. The two were drifting apart, so much so that they had to make a rule to take one night off a week to spend time with each other. Increasingly, these evenings became more strained and uneasy. Van Lawick was also beginning to chafe under Goodall's fame, declaring to one friend that he was tired of being "Mr. Goodall."[7] He was also no longer willing to sacrifice his photography to help his wife with the administration of the Gombe research center. Before things completely disintegrated, the two decided to divorce in 1974. Goodall remained on good terms with Van Lawick, but it clearly was a sad time for all, especially seven-year-old Grub.

Later on a visit to Paris to speak at a UNESCO conference, Goodall took the opportunity to visit the great cathedral at Notre Dame. Reflecting on her life and her failed marriage, Goodall realized she could go one of two ways. She could stumble through life without purpose or goals and succumb to the negative aspects of man: greed, selfishness, and anger. Or she could hold to the belief that she and her work were part of a larger whole that gave her life purpose and meaning. Goodall also recognized that as a result of her work at Gombe, she had undergone a spiritual awakening.

She acknowledged that there was a greater being at work and that human beings had a responsibility to the environment and its creatures. While some people would dismiss this by simply saying that all things in the end are the providence of God, Goodall held fast to the saying that "God helps those who help themselves."

Goodall's visit to Notre Dame sparked a call to action. Recalling that day, she believed she had heard the call of God. There was no question in her mind that she had undergone a powerful spiritual experience, one that brought back the intensity of her church-going youth in Bournemouth and her peaceful, introspective times in Gombe. Goodall later wrote that her visit was a turning point in her life. While she realized that she did not understand completely what had happened, she believed that when the time was right, God's will would be revealed.

A LOVE FOUND

After her divorce, Goodall threw herself into her work. As director of the research center, she came into contact more and more with various Tanzanian government officials. One of these officials was Derek Bryceson, Director of Parks and a member of the Tanzanian parliament. Bryceson had first visited Gombe in 1967 as Tanzanian minister of agriculture. In 1973, the two met again; Bryceson now was the new director of the Tanzanian national parks. In his new role, Bryceson was much more important to the Gombe research center. He was also more attentive to Goodall.

Bryceson was a tall, thin man, full of purpose and vigor. Born in 1922 in China, Bryceson was sent to England for his education. He left school in 1939 when World War II broke out to enlist in the British Royal Air Force as a fighter pilot. Bryceson carried a visible reminder of his service; he was in a terrible plane crash that left him with a severely injured spine. He was told by doctors that he would never walk again. Instead, Bryceson succeeded in teaching himself to walk with the help of a cane.

Bryceson then went to Cambridge where he earned a degree in agriculture. Then he set off for Kenya where he farmed for two years. He later applied for one of the wheat farms in Tanzania held by the British government. During that period, he met the compelling political leader Julius Nyerere. Bryceson soon became one of the earliest white supporters in the Tanzanian independence movement. Upon Nyerere's ascendance as president of the newly independent Tanzania, Bryceson was rewarded for his efforts through a series of important cabinet positions. In time, he emerged as one of the most influential members of Nyerere's government, and for many years was the only freely elected white person in all of postcolonial Africa. He was also deeply committed to the protection of the environment and animals.

When Goodall and Bryceson met in 1973, each was married to someone else. But that did not stop them from falling in love with each other. For Goodall, Bryceson was a hero, both as someone who had overcome great personal adversity and who had gracefully made the transition from British colonial rule to Tanzanian independence. He was also an idealist who shared a love of poetry and classical music with Goodall.

But the 1973 meeting was more about business than personal feelings. Goodall was especially anxious about the future of the research center. While at Stanford, she had been hopeful that her presence there would help raise funds for the center. But with the news of the Nixon administration's cutting back on federal funds for research in the sciences, Goodall had good reason to be concerned. She was also unhappy that the Tanzanian government, for some unknown reason, was delaying the clearances necessary for foreign students and researchers to come to Gombe. She took her concerns to Dar es Salaam, where Bryceson and other government officials lived. Making her case, Goodall asked for help. Bryceson and the other officials were impressed and for the time being, Gombe seemed safe.

Sometime during the summer of 1973, the two began an affair. Bryceson was a familiar visitor to Gombe and Goodall found reasons to come to Dar es Salaam. But the pair were apart more often than not; to keep in contact they maintained an intense correspondence for many months. Both were plagued, however, with feelings of guilt and uncertainty. Then, in early 1974, Goodall, Bryceson, and Grub were in a serious plane crash when Bryceson's small Cessna crash-landed. Goodall later described her last moments in the plane as preparing to die. Instead, she emerged from the burning plane with her son and lover and a new sense of enlightenment. For Goodall, more pieces of her purpose in life fell into place. Shortly thereafter, she accepted Bryceson's proposal of marriage. They wed in 1975, and Goodall moved to Dar es Salaam but still spent the majority of her time at Gombe.

A NIGHT OF TERROR

By 1975, there were as many as 20 students living and working at the research center. Graduate students from all over the world came to Gombe, representing a wide variety of interests and academic disciplines. Whenever possible, Bryceson came to Gombe, helping Goodall with the administration of the center.

In May 1975, the tranquility of the research center was shattered when 40 armed men crossed Lake Tanganyika from Zaire and raided the camp. The park warden, awakened by the shouting, went out to investigate. She was captured and with a gun held to her head was ordered to lead the intruders to the students' quarters. Despite her refusal to do so, the raiders had found four students and took them hostage. The warden and another student hurried to the students' quarters to warn them of the danger. Goodall, nursing an ear infection, was at her house on the lake and did not find out about the raid until after the men left the camp. She then gathered the other students and workers to try to decide what to do. The four students—three Americans and one Dutch—were still missing. Someone reported having heard gunshots by the lake. Goodall feared that the four had been executed.

The next several weeks were anguished and difficult. No word came about the fate of the students. In the meantime, the government ordered all foreigners to leave the research center. They were moved to Dar es Salaam to Bryceson's house and waited for news. About two weeks later, one of the students returned to Kigoma with a ransom demand. The intruders were revolutionaries who hoped to overthrow the government in Zaire. However, the demands were excessive not only in terms of ransom money and arms, but in the commitments imposed on Tanzanian government. Bryceson knew that the government either would not or could not seriously entertain the rebels' demands. Eventually, two of the rebels came to Dar es Salaam to negotiate with both the American and Dutch embassies. It was a torturous time, as people worried that the hostages would be killed.

Finally a ransom was paid. But counter to the rebels' promises that they would hand over the remaining three hostages, only two women were released. Goodall was in deep despair by this point, sure that the one remaining hostage would be killed. Finally, after several more weeks of waiting, the last hostage was sent to Kigoma. While the students were not physically harmed, they all had suffered deep psychological trauma.

It was a devastating blow for the research center. Gombe was now officially declared a "sensitive" area for many months to come. Visitors required clearance from the government to come. Without her husband's influence in the government, Goodall had little doubt that Gombe would have been shut down. Bryceson continued to help with the administration of the center, reorganizing the field staff to take on more of the responsibilities in gathering and organizing the field studies. The reorganization of the staff was a pivotal move, as no students would be allowed to return to Gombe until 1989.

RETURN TO STANFORD

Much to her dismay, upon Goodall's return to Stanford in 1975, she discovered that the kidnapping incident was far from over. To her surprise, she found out that the ransom monies paid to the rebels had, in fact, never been paid at all. Goodall was further shocked by the swirl of rumors surrounding her and her husband over the incident. Bryceson was accused of putting the students' lives in further jeopardy because he and the Tanzanian government did not want to initiate a dangerous precedent by giving in to terrorist demands. Further, Bryceson was willing to sacrifice the students so that his government would not have to pay. What the rumormongers did not know was that Bryceson, throughout the entire ordeal, met around the clock with government officials to work out alternatives. He had even gone so far as to contact the SAS, or Special Air Services, a British elite military unit that routinely undertakes special operations, to carry out a rescue mission. Goodall also came under criticism for what was seen as a lack of leadership. Critics charged that she should have taken the students' place as a hostage. Again, the fact that Goodall did not know about the kidnapping until after it had happened did not seem to matter.

At one point, Goodall was advised to leave Stanford at least until the controversy passed. She refused and instead ensconced herself in a rented house with Grub, Vanne, and her husband. Repeatedly, she confronted her accusers to explain her side of the story. The effort was costly and draining. But Vanne's presence was both a source of strength and comfort for Goodall. The two would plan strategies and share thoughts over what had happened and how Goodall could overcome the situation.

Distressed as she was, Goodall became even more depressed after an interview with a newspaper reporter who was investigating several high-profile kidnapping cases. After they had finished talking he told her that he understood why she felt her situation was shocking and unfair. But he told her that in every case he had studied, the exchange of large sums of money had also led to a breakdown in trust and friendship toward those who had suffered and given rise to bitterness and hostility. Goodall found his analysis disturbing and frightening. If he was right, it was a sad commentary on the state of the human condition. But Goodall was soon able to put her tough times behind her. By the end of the fall semester, 1975, she had fulfilled her teaching responsibilities at Stanford. She now looked forward to returning home to the peace and tranquility of Gombe. Little did she know that her fears about human behavior would be played out among her beloved chimps. There was much more darkness ahead.

NOTES

1. Sy Montgomery, *Walking with the Great Apes: Jane Goodall, Dian Fossey, Birutė Galdikas* (Boston: Houghton Mifflin, 1991), p. 86.

2. Jane Goodall, *Through a Window: 30 Years Observing the Gombe Chimpanzees* (London: Weidenfeld & Nicolson, 1987), p. 16.

3. "Jane Goodall," *New York Times Magazine*, February 18, 1973, p. 14.

4. Jane Goodall, *40 Years at Gombe* (New York: Stewart, Tabori & Chang, 1999), p. 30.

5. Sy Montgomery, *Walking with the Great Apes: Jane Goodall, Dian Fossey, Birutė Galdikas* (Boston: Houghton Mifflin, 1991), p. 43.

6. "Obituary" *London Sunday Times*, September 1972.

7. Sy Montgomery, *Walking with the Great Apes: Jane Goodall, Dian Fossey, Birutė Galdikas* (Boston: Houghton Mifflin, 1991), p. 125.

Chapter 9

MONEY, MURDER, AND MOURNING

As distressing as the recent events at Gombe were, Goodall turned her energies to continued research and writing. By now, Goodall had established herself as a significant and formidable presence in the field of primate studies. Her first book *The Shadow of Man*, published in 1971, described her early experiences and findings at Gombe. The book became an instant best-seller and was translated into several languages. The book is still in print as of 2004. Goodall had also written a children's book, *Grub, the Bush Baby*, based on her son in 1972. In addition, she wrote numerous scientific papers on her research, which was finally gaining a wider audience.

Goodall at first missed the energy and help that the students and researchers had provided. But in working with her African field staff, Goodall felt transported back to her first months at Gombe living by herself with only Hassam, Rashidi, and Dominic helping her. Still, research at Gombe continued with Goodall relying more and more on her skillful team of field researchers. Later, they would be joined by a handful of foreign researchers and a Ph.D. student from Tanzania. For Goodall, working with the Africans was important, for it allowed her to give something back to the community that had so willingly helped in her first days at Gombe.

Happily, Goodall now spent more time out in the field documenting chimp behavior. Whenever possible, she and Bryceson met, but the increasing demands as a member of Parliament and as director of the National Parks kept him away a good deal of time. To keep in touch, the two communicated by phone or field radio whenever possible. For both, it eased the loneliness they felt from being apart.

THE JANE GOODALL INSTITUTE

In spite of Goodall's growing reputation, the future of the Gombe Research Center was a troubled one. The termination of her position at Stanford diminished her chances of applying for more traditional grant monies that were usually given to people in academia. In some quarters, despite Goodall's graduate degree and work, there was still reluctance to give her money because she did not have a bachelor's degree. With the absence of a Ph.D. resident at the center, other funding dried up. Some grants did come through, which pumped more money into the center, such as an emergency grant from the Leakey Foundation. But the kidnappings had damaged the center's ability to raise money. With no students or researchers, foundations were reluctant to part with their money. In one case, a particularly large grant was not renewed, leaving the center in a more shaky position.

The National Geographic Society still provided funding, but these funds were often earmarked for *National Geographic* articles, lectures, and television specials about Goodall and her work. These projects clearly made money for the Society and raised Goodall's profile, but did little to maintain the day-to-day operations of the center. In desperation, Goodall stepped up her speaking engagements thanks to the backing and support of the Leakey Foundation. But chasing after money was tiring and frustrating, compounded by Goodall's homesickness for Gombe and the chimps.

In 1977, a significant financial breakthrough happened with the creation of the Jane Goodall Institute for Wildlife Research, Education, and Conservation in San Francisco, California. The Institute was helped into place by two of Goodall's closest friends, Prince Raniere di san Faustino and his wife Genevive. Faustino believed that the best way for Goodall to escape her money troubles was to create a nonprofit institute that would allow Goodall to build up her own funds for the Gombe research center. With his help, Goodall received legal status for a nonprofit institute bearing her name. With the creation of the institute, Goodall was no longer as dependent on grants or other outside funding to keep the research center going. Also helping Goodall were the supporters who gave their time and energy to help preserve not only the research at Gombe but to expand its programs. In time, other branches of the institute would be established in North America, Africa, Asia, and Europe.

THE KILLING TIME

By 1970 Goodall had amassed a wealth of information on chimpanzee behavior, much of it confirming her belief "that chimps were rather nicer than us."[1] Goodall also admitted that if she had stayed at Gombe for only a decade she would have continued to believe that chimps, despite their close genetic relationship with humans, were inherently good and incapable of dreadful acts of violence. So it was with a true sense of horror when Goodall learned that chimpanzees were not only capable of a primitive kind of warfare but even more shocking atrocities. The awful discoveries made during Goodall's second decade at Gombe shook her to the core and were particularly difficult for her to accept.

In early 1970, Goodall noticed that the community she had been observing over the last decade began to divide. Seven adult males, three mothers, and their children began spending more and more time in the southern region of the park. By 1972, Goodall noted that this group had completely broken away, forming their own community in the south; this community was now known as Kahama. The original community Goodall had been documenting, the Kasakela, continued to live in the north.

The relationship between the two groups was a strained one, as the Kasakela group was now being chased out of the southern area where it once had roamed freely. When males from both communities encountered each other in a neutral zone that overlapped the north and the south, one group would threaten the other. Usually the group with the fewer males at the moment would retreat and return home. At first, Goodall and her researchers saw no cause for alarm; the chimps were simply engaging in territorial behavior.

But over the course of the next few years, Goodall and her researchers began noticing an escalation in fighting whenever the two groups came into contact. The adult males of each group began engaging in increasingly violent behavior, shaking branches, throwing rocks, and charging at each other. Relations between the two communities continued to disintegrate, though the tensions and fights seemed still to be over territory.

In 1974, the unthinkable happened; the Kasakela males' aggressive behavior became even more violent toward the Kahama group. A village leader reported to Goodall that he had seen six Kasakela males moving south when they came upon a lone young Kahama male feeding. The chimps erupted into a frenzy, kicking, hitting, and biting the chimp almost to death. After the chimps left, the Kahama chimp, Godi, watched his attackers walk away. He was able to drag himself away, but was assumed to have died of his wounds, when researchers could no longer see or find him.

From 1974 to 1978, the chimp communities of Gombe engaged in what came to be known as the Four Year War, which resulted in the systematic annihilation of the Kahama community by the Kasakela group. All the attacks lasted anywhere from 10 to 20 minutes; in each case there were no survivors. Observers saw an additional four attacks on Kahama males in which there were also no survivors. In one grisly instance, the victim was found mutilated and brutalized; the remaining two males simply disappeared. The Kasakela females also were attacked; all five were either killed or went missing. By 1978, the entire community was gone, save three young females who were assimilated back into the Kasakela community. Goodall was devastated. She grappled with what was now another piece of the behavioral puzzle: chimpanzees, like their human counterparts, possessed a dark side to their nature. The realization changed her views about chimpanzee behavior forever. For months after the deaths of the Kahama chimps, Goodall struggled to come to terms with what she had witnessed and learned.

DARKNESS COMES TO PARADISE

While the Four Year War escalated, another shocking and horrific event unfolded at Gombe. In 1971, one of Goodall's researchers informed her that he had witnessed a terrible event. A group of Kasakela males attacked a female and her child; the female was beaten nearly to death, her child ripped from her arms, killed, and then partially eaten. The female managed to escape, but the researcher believed her wounds were mortal. The news stunned Goodall and the researchers at Gombe, leading to a discussion that ended in the early morning hours. As terrible as the incident was, Goodall believed that it was an isolated circumstance and would most likely not happen again. The chimp who led that attack was believed to be something of a psychopath who had preyed upon females in his own community. The only explanation for what had happened was that he somehow convinced the other chimps to join in the attack.

A similar attack occurred sometime later when again a group of Kasakela males attacked another chimp mother from another community. The mother and her infant were beaten so severely that both died. The senseless attacks troubled Goodall. She also realized that the idea of "the noble ape" possessed of a more peaceful nature and incapable of committing harmful acts was truly a myth. If anything, the chimps were demonstrating violent tendencies that were more in keeping with man's own aggressive and violent behavior.

The worst was yet to come. In 1975, shortly after the kidnapping of the four students, Goodall received a very disturbing message from Gombe: a researcher had documented a number of cannibalistic attacks made by the dominant female, Passion, at Kasakela against other females in her own community. This was unheard of; Goodall and Bryceson, listening to the report, could only look at each other and ask why. The two later flew to Gombe from their home in Dar es Salaam to get a briefing on the matter.

According to one of the field assistants, Gilka, another female that Goodall had watched from infanthood, was cradling her child when Passion appeared, with a daughter, Pom. Passion was an older female chimp. Unlike Flo and her daughter Fifi, who were kind and caring mothers, Passion showed little interest in Pom as an infant and child. She often left the child alone to fend for itself. As Pom grew older, however, she spent more and more time with her mother; from all aspects, the two now seemed to get along.

According to the researcher, Passion stared at Gilka for a moment then charged at her, her hair bristling. Gilka screamed, afraid to defend herself; she had earlier been crippled in the wrist from the polio epidemic. Her infirmity and attempt to protect her child put her at Passion's mercy. Passion seized the baby boy and with one strong bite to the baby's head, killed him. Accompanied by her daughter and her son, Passion then sat down and promptly began eating the child. Gilka by this time had fled.

As horrible as the attack was, what depressed Goodall most was the cannibalism. Again, she and her researchers met to make some sense of the situation. Why had Passion attacked Gilka? Clearly, Passion was not in need of food, nor was Gilka from a neighboring chimp community. The two females had known each other all their lives. There was no rational explanation for what had happened.

Goodall now wondered if Gilka's first baby, born a year earlier, had met with the same grisly fate. The infant had vanished mysteriously a few weeks after its birth; it had been assumed that the child had died from sickness or an accident. But now, given the recent event, it was possible that Passion had killed Gilka's first child, too. Goodall's worst fears about Passion were confirmed a year later when Gilka again gave birth. Passion killed the infant. This time Gilka put up a fight; but while Gilka was fighting Passion, Passion's daughter, Pom, seized the child and killed him. Mother and daughter then ate the child. Gilka was severely injured and her physical wounds never completely healed. But Goodall also believed that the loss of three children had left Gilka broken; what spirit and energy she had was gone.

Two years later, Goodall discovered Gilka dead by the Kakombe stream where Flo had died a few years before. Goodall reflected on how sad Gilka's life had become. The once bouncy and gay little girl had become a victim: first of polio, later of a severe fungal disease that had cruelly deformed her, and finally of a series of vicious attacks that injured her and claimed the lives of her children. After the death of her mother, Gilka appeared very lonely, but an elder brother took care of her and watched over her when they traveled together. When Gilka gave birth to her first child, Goodall sensed through Gilka's handling and affection for the child, that she would be a very good mother. Looking down at Gilka's body, Goodall felt a sense of peace; Gilka was at last free of a life that had become burdensome.

Against the backdrop of the Four Year War, 10 infant chimps were born in the study community at Gombe. Only one child, born to Fifi, the daughter of Flo, survived. The researchers knew that Passion and Pom had killed and eaten at least five of the infants; Goodall believed that the remaining three were also victims of Passion and Pom. But she noted that without her daughter, Passion could not carry out an attack against another female chimp and take her child from her. In 1976, two African staff members recorded another attack by Passion and Pom on Melissa, a gentle-natured chimp that Goodall had closely documented. The staff members threw rocks at the two chimps in the hope of driving them away. Pom ran but not before killing the child and taking it away. Then Passion went over to Melissa, touched her wounds, and embraced her. As Goodall later interpreted these actions, the embrace was Passion's way of telling Melissa that she had no quarrel with her, but only wanted her baby.

At the center, Goodall and her researchers discussed ways in which other attacks could be prevented. In general, the staff and Goodall thought very carefully before they stepped in. But before anything could be done, Passion and Pom both gave birth to their own children. The killing and cannibalism stopped. Goodall named Passion's baby Pax—Latin for "peace." But Goodall still struggled to make sense of these barbaric attacks. In an interview years later, Goodall acknowledged that the infanticide "was the hardest thing to understand and accept that's ever happened at Gombe."[2]

In 1979, Goodall published her findings in an article for *National Geographic*. The article, "Life and Death at Gombe," again engulfed Goodall in controversy. Both she and the magazine were criticized for publishing the article, citing it as gruesome and exceptionally graphic in its portrayal of the violence at Gombe. But the worst was yet to come, as Goodall later wrote: "Some critics said that the observations were merely 'anecdotal' and should therefore be disregarded. This was patently absurd. We had watched, at close range, not just one but five brutal attacks.... Even more significantly, other field researchers had observed similar aggressive territorial behavior in other parts of ... Africa."[3] Goodall was further angered by the attitude of other scientists who took the behaviors of the Gombe chimpanzees to theorize not about the roots of chimpanzee aggression, but to substantiate or refute their own theories on the nature of *human* aggression.

Two years later, Goodall was asked to give a lecture at a Leakey Foundation fundraiser. She reluctantly agreed. The topic of the talk was "Cannibalism and Warfare in Chimp Society." Before she began her talk, she apologized to the audience, stating the subject had been picked for her. Given her own choice of material, she would have much rather emphasized the more positive aspects of chimp behavior. What stands out from this lecture and the earlier article was that Goodall had successfully proven an important part of her research: that an animal's individual temperament, family background, and decision making were all rooted in the chimp's own personal history. In her talk, Goodall also pointed out that the actions of one or two chimps can affect the whole community; for instance the actions of Pom and Passion affected an entire generation. The attacks on mothers and their children stopped the raising of all but one infant in Kasakela. That these horrible events were played out against the terrorist raid on Gombe in 1975 led Goodall to one important observation: Chimpanzees were far more like humans than she originally thought. It was a chilling conclusion.

A LOST LOVE

Amid the terrible events that plagued Gombe, Goodall's star was ascending. Between 1974 and 1984, she received a number of prestigious awards including the J. Paul Getty Wildlife Conservation Prize, which presented Goodall with the sum of $50,000 for her work. She was the subject of numerous articles, and the National Geographic Society continued to produce television specials documenting her work. In an ironic turn of events, Goodall, in the midst of all the publicity about her work at Gombe, was actually spending less and less time there. After the 1975 kidnappings, the government monitored her visits carefully, even restricting the time she could spend there.

When not speaking abroad, Goodall now spent more time with her new husband. Those who saw the couple believed that Goodall had truly found the love of her life. When Bryceson visited Gombe, many found him to be a more easygoing presence, compared to the intense, chain-smoking Van Lawick. For Goodall, life in general had become more relaxed.

She was in love and happy, even with the horrible events at Gombe in the background.

But her happiness was cut short. In 1979, Bryceson began complaining of stomach pains. Upon discovery of an abdominal mass, the two then flew to England to see another doctor. The doctor told Bryceson that there was a tumor present, but he believed that the growth could be removed with surgery. The night after her husband's surgery, Goodall was told that her husband's condition was far worse than earlier thought; Bryceson was in the final stages of cancer. At the most he had three months left to live. Once again, Goodall drew strength from Vanne's presence and support.

The two decided to try alternative types of medicine, but there was nothing available in London. Desperate, the couple went to Germany. For the first two months, it appeared that Bryceson was improving. He began working on his autobiography; he and Goodall talked and listed to classical music. Friends from England and Tanzania called and visited. The Tanzanian ambassador to Germany was also a frequent caller.

During her husband's convalescence, Goodall stayed in a small hotel nearby. On her way to the hospital she often picked wildflowers as a gift for her husband. But by the third month of treatment, Bryceson's condition worsened. Goodall could not leave him, often spending the night in a small bed that the hospital staff provided for her. Night after night Goodall watched as her husband fell further into a coma. On the night of October 12, 1980, Goodall lay awake and listened to his labored breathing. Earlier that day, he did not regain consciousness; Goodall knew the end was near. She heard her husband take one last breath and then heard his death rattle. Goodall knew that Bryceson was now free of the pain and suffering he had endured over the course of the last year. She climbed into his bed. When the nurse arrived she found Goodall still holding her husband close. After only five years of marriage, her companion and soul mate, the great love of her life was gone.

Bryceson's death was a wrenching blow to Goodall. After his death, she returned to the Birches for a short time, and then went to Dar es Salaam. In November, during a heavy rain, a memorial service was held for her husband in a small building near their home at the lake where the two had often swam and explored. After the service, the mourners gathered into three boats and headed out to a favorite spot that the two often visited to see the coral reef. As she watched, Bryceson's gray ashes were cast onto the waters. As the service drew to a close, Goodall watched the ashes float away. She returned to the house she had shared with her husband, alone. In later years, Goodall recalled her husband's illness and death as one of the cruelest and darkest times of her life.

A FAITH TESTED

For the next several months, Goodall quietly grieved. True to her character, she talked little of her husband's death to family and friends. She continued to wear her wedding ring; it can still be seen on her hand today. Her grief was compounded by two more incidents: the death of her grandmother Danny Nutt and the emergency surgery of her mother for a heart problem. Vanne did recover, but the deaths of Bryceson and Danny only months apart pitched Goodall into a darkness she had never known.

The cruel manner in which Bryceson had been taken from her sparked a deep spiritual crisis for Goodall. During their time in Germany, the two had often prayed together in the hope that the cancer would be vanquished. By now she felt as if God had deserted her. She had tried everything she could to save her husband's life, including folk medicine and trips to an Egyptian faith healer and an Indian spiritualist. She had prayed religiously every day for her husband to be cured, but God had not listened to her entreaties. He had deserted her.

Goodall then became angry: at herself, the doctors, anyone who had remotely been a part of Bryceson's illness. Most of all she was angry with God and what she believed was the unjust way in which she and Bryceson had been treated. For the next several months, Goodall withdrew from the spiritual comfort and solace that had helped her so much in the past. Turning her back on her faith only increased the bleakness and despair of her life.

Searching for peace of mind, Goodall returned to Gombe shortly after the funeral. When she arrived at the center, she was greeted by staff and workers all offering their condolences for "Mr. B." Goodall knew that many were now concerned with their own future and that of the park and center. But all Goodall wanted was to seek out the chimps in the hope that they would help her with her grief.

Her first days at Gombe were painful. The house, once the source of so many happy memories, was now filled with ghosts. Then on the third day, Goodall decided to hike to the feeding station. At one point on the trail she felt herself smiling; she was on the part of the trail that her husband with his crippled limb found so difficult to climb. But now it was Goodall who was struggling. She could feel Bryceson light and free, teasing her. She laughed aloud, for the first time in weeks. Later that night Goodall was awakened from her sleep. She saw her husband standing before her. He spoke to her about things she needed to know and things she needed to do. When Goodall woke up, she knew her husband had come to her and had imparted her next mission. For months afterward, Goodall often felt his presence: "Perhaps it was fancy, but it comforted me, the thought that he was there, that I could do something for him. And then, after a while ... I felt his presence less and less often. I knew it was time for him to move on and I did not try to call him back."[4] Goodall had also found her way back to her spiritual center and once again began working. To honor the memory of her husband, Goodall embarked on some new projects. The first was the creation of the Bryceson Scholarship Fund. With seed money given to her, Goodall established the fund that grew thanks to donations from individuals and fund-raising lectures. The scholarship was important not only in a financial sense but in a symbolic way; it demonstrated again Bryceson's commitment to a better future for his adopted country.

CHIMPANZOO

In 1984, Goodall embarked on another project in honor of Bryceson: the Chimpanzoo. Goodall had recently returned from a speaking tour during which she visited a zoo. Struck by the poor conditions of the chimp area, Goodall wondered what might be done to improve them. Eventually, a Gombe research student, Ann Pierce, surveyed American zoos and created an organization to coordinate a long-term study of behavior in zoo chimpanzees to help better the conditions in which they lived.

Helped in part by the Jane Goodall Institute, Pierce set out in an old van to travel the country. Soon she had set up collection boxes at a half-dozen new zoos. Pierce also developed contacts with universities and colleges that pledged to send student volunteers to collect the data. Their findings would be used to create a database that eventually would be computerized.

The project was a huge success; besides the database, which includes information on approximately 130 chimpanzees, annual conferences were held to discuss research. Chimpanzoo is still thriving thanks in large part to the support of the Jane Goodall Institute, and has expanded to publish newsletters, program reports, monthly reports, and establish site visits and annual conferences.

RESURGENCE AT GOMBE

While overseeing the scholarship fund and Chimpanzoo, Goodall also returned to her home and spiritual center: Gombe. After Bryceson's death, the bans on visitors and students coming to the park and the research center were lifted. From 1981 to 1986, the center was a whirlwind of activity.

Goodall, still acting as administrator, now oversaw the monitoring of the chimps and the record keeping done by her talented and competent African staff. The Tanzanian field assistants and staff began videotaping the chimps to create a visual record of the research at Gombe. One of the first events recorded was of a mother chimp giving birth.

Growing numbers of researchers and scientists also came to Gombe to start on projects of their own or to meet with Goodall and discuss research. One anthropologist from the University of Northern Kentucky was Christopher Boehm, noted for his work in vocal communications with chimpanzees. Goodall met with other noted primatologists, who, in turn, often invited her to come to their universities to teach or speak. During this period, Goodall also reached out to Japanese primatologists in the hope of trading information and research. In 1982, she was recognized by the Japanese government with the presentation of the Kyoto Prize for Basic Sciences. The award is Japan's highest honor and is considered to be the equal of the Swedish Nobel Prize. After the award ceremony, Goodall spent time in the mountains of Japan with the pioneer of Japanese primate studies, Junichero Itani and his chimpanzees.

The research center also underwent a facelift. The student dining hall was transformed into guest rooms for tourists. Park officials and their families moved into the tin buildings that housed the African staff. Though crowded, it added a growing presence of Tanzanians who now held positions of all kinds. Even though tour operators are permitted to bring white tourists to see Gombe, white researchers and students are still prevented from working at the site except under special permission from the Tanzanian government.

THE CHIMPANZEES OF GOMBE

With all of the activity surrounding her, Goodall embarked on the most draining and time consuming project of all: a book, *The Chimpanzees of Gombe: Patterns of Behavior*. She had been at work on the book since the late 1970s. At the time, she envisioned the final product as not more than an update of her 1968 monograph *The Behavior of Free-Li Chimpanzees in the Gombe Stream Reserve*.

Since 1968, however, the center had accumulated a vast amount data. When Goodall began her research in 1960, she was armed or with a pair of binoculars and her field notes. But with the creation of th research center, many more people had stepped in to help collect data By the time Goodall started writing in earnest, it was estimated that she had approximately 80 years worth of research to draw from. The data was found in bursting file cabinets, cartons, and boxes at the Dar es Salaam residence, with more piling up on every available surface in the house. The task ahead of her was daunting: Goodall would now have to sift through the mounds of data, synthesize it, and make some sort of sense out of it.

During her writing, Goodall was aided by a number of research assistants. But even with their help, Goodall struggled not only with her writing, but in fighting off continual bouts of malaria. The deteriorating political situation and growing unrest was also cause for concern. Tanzania's experiment in socialism had failed miserably, causing an economic collapse with shortages, inflation, unemployment, corruption, and rising crime. Goodall's house at one point was broken into—no one was harmed and nothing was taken, but it was a grim reminder of how dangerous conditions had become.

Still, by 1983, Goodall wrote to friends and family that the book was almost done. Once completed, the book was an enormous (650 oversized pages) and extraordinarily complex work. The book consisted of 19 chapters; sections were color coded with dozens of intricate maps and drawings scattered throughout. In addition, there were several hundred photographs, 5 appendices, 20 pages of references, and 2 indexes. In its completed form, *The Chimpanzees of Gombe* was the most thorough work of its kind, touching on most every aspect of chimpanzee behavior.

The book finally rolled off the Harvard University presses in the summer of 1986 and appeared in bookstores that fall. The book was truly monumental; it explained Goodall's study of chimpanzee behavior. The book was a critical success and went through several reprints. It not only garnered favorable reviews but was awarded the R.R. Hawkins Prize for the Most Outstanding Technical, Scientific, or Medical Book of 1986. The book was the focus of a major international conference of chimpanzee experts sponsored by the Chicago Academy of Sciences. Almost 20 years after its publication, *The Chimpanzees of Gombe* remains the most comprehensive work on chimpanzee behavior ever published. It is the bible for many in primate studies, expanding the research of chimp behavior all around the world.

The Chimpanzees of Gombe signified a crowning moment in Goodall's career. For the last decade she had suffered through death, tragedy, and the fear that the Gombe research center would be closed forever. Many of her earlier observations and beliefs about chimp behavior had been shattered due to the violence and terror of Four Year War and the infanticide. She had overcome grief and a deep spiritual crisis, only to emerge stronger and more committed to the chimpanzee than ever before. Now, after almost 26 years in the field, she was finally receiving public acknowledgement for her pioneering efforts and achievements in the study of man's closest relative. For the next two decades, Goodall expanded her horizons, focusing not only on the chimps of Gombe, but also on the plight of chimpanzees everywhere.

NOTES

1. "Jane Goodall," *Current Biography Yearbook, 1991* (New York: H. W. Wilson Company 1992), p. 251.

2. Sy Montgomery, *Walking with the Great Apes: Jane Goodall, Dian Fossey, Birutė Galdikas* (Boston: Houghton Mifflin, 1991), p. 122.

3. Susan McCarthy, "Jane Goodall: The Hopeful Messenger," October 27, 1999, http://www.salon.com/people/feature/1999/10/27/reason (accessed October 15, 2004).

4. Susan McCarthy, "Jane Goodall: The Hopeful Messenger," October 27, 1999, http://www.salon.com/people/feature/1999/10/27/reason (accessed October 15, 2004).

Chapter 10

THE CELEBRITY AND THE CRUSADER

With the publication of *The Chimpanzees of Gombe* in 1986, Goodall's life took another dramatic turn. The 1986 conference, "Understanding Chimpanzees," not only was celebrating the publication of Goodall's masterful work, it was also one of the largest gatherings of chimpanzee experts from around the globe. Attending the three-day conference were researchers, scientists, and contemporaries of Goodall. Also attending were former Gombe students, zoo officials, and representatives from other research facilities and laboratories. The scientists were carefully screened; only those who did noninvasive work were permitted to attend.

A STARTLING REVELATION

During the final sessions, the discussion began moving away from research and toward the state of chimpanzees in general. The question was raised again and again that while research was important, what would happen if there were no chimps left to study? For many, the question had two parts. Wild chimpanzees in Africa were in serious danger from the degradation of their habitat, the threat of hunters, and the capture of infant chimps for sale. Of concern, too, were the living conditions and health in zoos, labs, private homes, and the entertainment industry.

At the conclusion of the conference, 30 top experts in the field of chimpanzee research decided to establish an organization to help preserve the wild chimpanzee and to create standards of care for chimps in captivity. The new organization, the Committee for the Conservation and Care of Chimpanzees (CCCC), was promised financial support from the Jane Goodall Institute. However, the organization would work independently. The CCCC promised to be an important and formidable voice to speak on the behalf of the chimpanzees. The chairman, Geza Teleki, a former Gombe student, was a strong personality with sharp political instincts, who many hoped would get the job done.

For Goodall, the conference was a revelation. She had come to Chicago as a chimpanzee researcher to talk about her book. She was already planning to write a follow-up volume. But as she said in a later interview, something quite different happened to her: "Slide after slide, presentation after presentation, brought home how bad things were—encroachment into reserves, deforestation, habitat loss, poaching, chimps being snared.... It was a real road to Damascus moment for me."[1] Goodall entered the conference as a researcher and supporter of the Gombe chimps. She left a committed activist and reformer, dedicating herself to the protection and care of all primates and to the world they lived in.

A CHANGING PERSPECTIVE

Goodall's remark about the road to Damascus was revealing. The reference was biblical, describing the journey of Saul of Tarsus, who persecuted the early Christians. On his way to Damascus, Saul was struck blind by God; he then converted to Christianity, took the name Paul, and for the remainder of his life, carried the gospel to others. Today, he is regarded as one of the most important disciples of Christ.

Certainly, Goodall could not be charged with being inactive; she had expended considerable time and effort to educate people about the chimps of Gombe. But her emphasis was only on Gombe. She later admitted that she had isolated herself in her own little world. Between her work and raising Grub, she failed to see the larger picture unfolding before her. She knew that around Gombe the habitat was beginning to disappear, but the park itself was safe from development, hunters, poachers, and other threats to the animals. What she heard at the conference was more than shocking; it was a call to action.

Before her she saw what was happening in other places in the world: the bonobo or pygmy chimp resided in a war zone in Zaire. Orangutans in Asia were losing their habitat due to destruction of the forestland by terrible fires. In central Africa, their last great stronghold, chimpanzees were being hunted commercially for food. Logging companies began opening up the forest, hunters with high-powered weapons now had only to go along a road, riding on a truck and everything—chimps, gorillas, and bonobos—was in plain sight. The carcasses were stripped for the meat, which was dried or smoked and sold, not to people who needed it, but to those who found bush meat a delicacy. Hands and feet of the monkeys and chimps were sold as novelty items, or ground up for potions to bring strength and courage. Goodall also read reports that if the current trends were not reversed and some kind of agreement was reached with the African governments, hunters, poachers, and logging companies, that within two decades there would be no apes left in the great Congo Basin and even fewer throughout the continent.

A NARROW VIEWPOINT

The news that Goodall was now committing herself to a much larger focus was greeted warmly. For years, animal rights activists had begged Goodall to use her celebrity to join them in their cause. They told her that while the chimps at Gombe were safe, there were thousands of others who were not. But Goodall was not interested at the time and continued to concentrate on her chimps at the reserve.

Geza Teleki, who had worked at Gombe in the 1960s, later went on to establish a national park in Sierra Leone, West Africa. Before his efforts, chimps were shot for food or were captured and sent to labs for medical research. Teleki also knew that for every chimp sent overseas to a buyer, anywhere from 5 to 10 died in transit. Hunters who could not afford bullets sometimes fired buckshot made from metal shards. They not only killed their intended targets, usually female chimps, but often the infant originally intended for capture.

Teleki remembered how at Gombe Goodall worried and fretted over the fate of individual chimps. She'd suffered through the polio and flu epidemics. She grieved over the deaths of chimps such as David Greybeard and Flo, whom she had come to regard as close companions; when necessary she had taken matters into her own hands to put suffering chimps such as Mr. McGregor out of their misery. So it came as a surprise to Teleki that when told about the plight of the chimps at Sierra Leone, Goodall had little empathy for them.

For Teleki, her indifference was devastating and infuriating. By this time, Goodall's high profile would have helped Teleki and others to save chimps. He believed that Goodall had a responsibility to speak out not just for chimps at Gombe, but for chimpanzees everywhere. But what Teleki and other animal rights activists discovered was that Goodall was more oriented toward individuals and personalities. Her narrow-minded focus kept her from looking at the larger problem at hand. Instead, she chose to concentrate on her chimps at Gombe.

The fact that Goodall continually refused to speak out about large-scale chimpanzee preservation left other animals rights activists both puzzled and irritated. One conservation activist echoed Teleki: Goodall was popular and could have made chimp preservation a hot-button issue. Instead, she chose not to speak out publicly and refused to lend her name or testimony to efforts to raise money or garner publicity. She also refused to speak to politicians about lobbying for better treatment of captive chimpanzees. Even as Dian Fossey fought off poachers and pestered politicians to protect the gorillas in Rwanda throughout the 1970s and 1980s, Goodall would not make a fuss. Her chimps and Gombe were safe and that was all that mattered. As her critics pointed out, Gombe, despite her efforts, was still under threat from poachers and development, but Goodall continued to look the other way.

"I OWE IT TO THE CHIMPS"

By the end of 1986, however, Goodall had thrown herself completely into this new cause. As one friend commented: "It was as if Jane suddenly got religion."[2] With the success of *The Chimps of Gombe*, Goodall now had the self-confidence to speak out to medical researchers, scientists, and politicians about the growing threat to chimpanzees. She embarked on a series of speaking engagements that increasingly kept her away from Gombe. She traveled in Africa with an exhibit called "Understanding Chimpanzees." Goodall met with heads of state and environment and wildlife officials. She helped push for Wildlife Awareness Weeks, in which she visited schools and libraries, attended fund-raisers, and gave public lectures. She made as many media appearances as she could.

In the United States, Goodall worked closely with the CCCC, which published extensive documentation about the conditions of wild and captive chimpanzees. The organization wished to list chimps as an endangered species to afford them international protection. They also pushed for assurances that labs and zoos would be more conscious of the physical and mental health of the chimps in their care. Goodall took the documents and lobbied legislators and other Washington officials. She gave countless interviews and called press conferences, then made the rounds of television shows such as "Nightline," "Good Morning America," and "National Geographic Explorer." Her lectures focused on her new agenda. Her trips to Washington, London, and Dar es Salaam increased from 1 or 2 a year to 8 or 10. When asked why she was working so hard and sacrificing her time to be with the chimps at Gombe, Goodall simply stated, "I owe it to the chimps."[3]

In spite of her hard work, some animal rights activists were still angry that Goodall waited so long to join the battle. Goodall preferred not to analyze too deeply her reasons for not speaking out earlier. She does admit, though, that while living and working in Gombe she became isolated and selfish in paying attention to the Gombe chimps. She also said that perhaps she should feel guilty about not helping sooner, as the welfare of all animals has been of vital concern to her since she was a small child. Whatever the reasons, Goodall had at last committed herself to a much wider stage and role than ever before.

NIGHTMARE IN THE LABORATORY

In December 1986, a group of radical animal rights activists secretly entered an ordinary looking building owned by a company called SEMA. The facility was a medical research laboratory that housed around five hundred primates including chimpanzees and monkeys. The animals were being used to test experimental vaccines for respiratory and hepatitis viruses, as well as the deadly AIDS virus. The animals were monitored to see how the vaccine worked or the disease progressed. Many primates died slow and horrible deaths from these experiments.

The intruders photocopied records and stole two cages and four chimpanzees. They also made some rough videotapes of the conditions inside the lab. The videos and documents were then sent to a variety of animal rights organizations including PETA (People for the Ethical Treatment of Animals). PETA edited the tapes, producing a short, 18-minute video called *Breaking Barriers*. The video and other documents were sent to people PETA believed might be interested. One of the recipients was Jane Goodall.

Goodall was visiting the Birches for the Christmas holidays when PETA's package arrived. She later sat down with her family and watched the tape. What she saw in the brief video left her in shock, anger, and sadness. For someone who had spent most of her life working with wild chimps, what she now witnessed was something out of hell. In early 1987, Goodall asked and received permission to visit the SEMA facility. What she saw was chilling; not only did the lab have an exceptionally high mortality rate, but the conditions under which the chimps and monkeys lived were abominable. The animals were placed two in a cage that left them little room to move. Once injected with the research viruses, the animals were isolated in a closed steel box with a small window. The animals were kept in isolation for as long as two to three years. For Goodall, the conditions were not only simple cruelty but were obscene.

When Goodall confronted the head of the facility about the conditions, he replied that the conditions were acceptable and nothing needed to be changed. Goodall left, vowing to do something about that and other, similar facilities.

The visit to SEMA led to more visits to labs and other medical research facilities all around the world. On her speaking engagements, she graphically described the lab conditions she had viewed, urging her audiences to get involved and push for the more humane treatment of animals used for lab research. She organized conferences through the Jane Goodall Institute that brought together chimpanzee and primate experts who discussed ways to improve the conditions in labs. Her relentless pursuit of better conditions led the United States government to amend existing regulations, emphasizing that labs concentrate on performance rather than engineering standards. This simply meant that the environment and treatment of all lab animals had to show improvement in order to continue research and testing. Even now, as she travels and visits these same facilities, she talks to their officials and offers suggestions on how the labs might improve treatment, even asking if they would stop using primates and other animals for experiments.

OTHER CAMPAIGNS

Throughout the remainder of the 1980s, Goodall lobbied for other changes. Along with Geza Teleki and the CCCC, Goodall pressed for a change in the U.S. Endangered Species Act. In 1975, the Convention on International Trade in Endangered Species of Wild Fauna and Flora (CITES) was created. The United States was one of the first 10 nations to ratify it, putting enforcement of the treaty under the Endangered Species Act, created in 1973. Chimpanzees were listed in Appendix I of the CITES treaty, signifying that they were in danger of extinction. This meant they could not be traded between countries if the motive is primarily commercial, or if taking them from the wild would threaten the survival of the species.

In 1987, a report written by the CCCC and supported by Goodall formally requested the U.S. Fish and Wildlife Service upgrade the status of chimpanzees from "endangered" to "threatened." The problem was that the act had little impact on the comings and goings of chimps within national borders, particularly those animals being transported for research and experimentation. But the act did restrain the international trade of chimpanzees from Africa. If a federal or federally funded organization was to trade in chimps, they had to apply for a permit and demonstrate that the importation would not threaten the continued existence of the animals. The U.S. Fish and Wildlife department then examined the petition and from March 23 to July 21, 1988 provided the request for public commentary.

The response was overwhelming; more than 54,000 postcards, letters, and other messages poured into the department's mailbox supporting the request. In addition, more than 40 responses were recorded from African governments, other organizations, and experts who had studied chimpanzees. It was the largest public participation on the issue of endangered species ever.

There were letters of protest as well, ranging from circus owners to officials with several large biomedical research laboratories. In the end, the U.S. Fish and Wildlife Service offered a compromise: chimpanzees would be defined as "endangered" if they lived in Africa, outside of the continent, chimps would be classified as "threatened." The compromise also created a loophole that allowed for the transport and sale between states for the purposes of medical research testing. While not the response they hoped for, Goodall and other animal activists had certainly raised awareness of the problem and had, in the process, gained more supporters for their cause.

MISSING GOMBE

Goodall now rarely came to Gombe. When she did, it was often sandwiched in between speaking engagements or events. Still, through writing, photography, and video, she continued to record the world and behavior of the chimpanzees who lived there.

In the closing years of the 1980s and early 1990s the Gombe research center documented a number of events—happy, exciting, and sad. In 1987, an outbreak of pneumonia afflicted the community, killing 11 chimps. It was a worse epidemic than the polio outbreak two decades earlier. In the outbreak, a small male chimp, Mel, lost his mother. Orphaned and frightened, Mel was lost, not knowing what to do. Many thought the young chimp would die.

To the amazement of the researchers, a young male chimp, Spindle, adopted Mel, and took over his care and upbringing while protecting him from the other chimps. It was the first recorded adoption by an unrelated chimp of an orphaned youngster. It also demonstrated to observers that males could be just as good caretakers as females. At the risk of irritating the male adults, Spindle waited for Mel when traveling, carried him on his back, shared his food, and let Mel creep into his nest at night. He was affectionate and nurturing toward Mel and taught him the needed skills to survive. There was no doubt in anyone's mind that Spindle, through his altruistic actions, had saved Mel's life.

In 1993, researchers also began studying another new community, the elusive Mitumba, which resided in the south, near the same area where the doomed Kahama community had once wandered. The group provided observers a chance to study the birth and upbringing of the second set of twins ever born at Gombe, Roots and Shoots, the infant sons of Rafiki, in 1995. But once again tragedy struck the small community when a severe respiratory illness claimed eight members, including Rafiki and her twins. It was a devastating loss for the community, which had numbered only 29 members.

In another first, one member of the Mitumba community, a female, joined the Kasakela group. Observers then noticed that Flossi, the daughter of Fifi, was using the Mitumba technique to capture carpenter ants with twigs. It was the first documented instance of a "technology transfer" from one chimp community to another.

There have been other findings as well: when thirsty, chimps take leaves and squeeze them like a sponge for water. They can strip a branch of its leaves to use it as a digging stick or as a hook to gather fruit. To crack open nuts, chimps use a rock and a hard platform. Chimps learn tool use from each other. They keep their nests clean, making sure all waste is disposed over the edge. Perhaps most important of all, chimps are social creatures, with a wide range of emotions, feelings, and behaviors. For Goodall, these findings have been important, for they helped extend researchers' knowledge of chimp behavior. It also helped elevate Gombe as one of the longest-running and distinguished research facilities in the world.

ROOTS & SHOOTS

Goodall continually tried to think of ways to educate more people about the preservation of chimpanzees and other primates and the conservation of natural resources. One audience she quickly tapped into was children. Many children were already familiar with Goodall through her children's books. But Goodall needed to find a way in which children from all over the world could become more involved with animal and environmental conservation.

In 1991, while meeting with a group of students on her front porch in Dar es Salaam, Goodall conceived her next effort: the Roots & Shoots program. She saw how the students were fascinated by animal behavior and how concerned they were about the state of the environment. They told Goodall that they wished to learn more, but that their schools didn't cover these topics and could not provide other resources that might help them. The students asked Goodall if there was a way they could pursue their interests through out-of-school activities. They talked about possible projects, including how to help chimpanzees and how efforts to save the environment might affect their communities.

Goodall sent the students back to their schools with an assignment: they must find other interested young people and take action. With the creation of the Roots & Shoots program, groups of school children and teenagers carry out small-scale, local projects that benefit their community, animals, and the environment as a whole. The program was an immediate success.

By 1993, Roots & Shoots arrived in America. In January of 1994 there were 11 registered groups in North America. Today there are close to 1,100. Goodall also began organizing annual summits for college and university students and achievement awards for groups and individuals. The organization published a newsletter that helped promote the exchange of ideas and research among Roots & Shoots groups throughout the world.

Goodall found the response overwhelming and gratifying. To see young people gather together to discuss what they could do to make the world a safer place gave her hope that the future might be a little better because of their efforts. She was also amazed at some of the projects the young people designed and cheered them in their successes.

When Goodall traveled to visit Roots & Shoots groups, she saw how the program not only brought people together but also how some projects garnered public acclaim. Visiting a Roots & Shoots group in Israel, she saw that Israeli students had joined together with Palestinians. The two were organizing a project together. Five children who joined their primary school group tackled a baboon problem in Musoma, Tanzania. Baboons had been raiding crops, so villagers were burning their forest in retaliation—which gave the baboons less food and made them even more aggressive.

To help solve the problem, the students gave talks to villagers on the importance of the baboons. They then planted crops and food in the forest for the animals to prevent them from coming into the village to steal the crops. Within three years the raiding had virtually stopped. Goodall was even prouder when she learned that the five children, who lived in terrible poverty, were invited to Sweden to be presented with a conservation award from the Volvo Automobile Company. Since its inception, Roots & Shoots has spread rapidly. More than 6,000 groups—ranging in size from 2 to 2,000—have registered in more than 87 countries. It continues to be one of the most successful programs that Goodall has created.

TACARE

In 1994, the Jane Goodall Institute initiated the TACARE project with funding from the European Union. TACARE, which stands for Lake Tanganyika Catchment Reforestation and Education project, was designed as a pilot program to seek ways of arresting the rapid degradation of natural resources, especially the remaining forestland in the Kigoma region of Tanzania. At that time, the Kigoma region had the second-highest yearly deforestation rate in the country. The rapid growth of the local populations plus the addition of thousands of refugees from the war-torn Burundi and the Democratic Republic of Congo placed a severe strain on the natural resources of the region. The TACARE project focuses on community socio-economic development by offering training and education in resource management. Goodall believed that involving local people in conservation projects would not only help protect chimpanzees and other endangered species, but would also bring a measure of economic relief as well as sending a message that they and their problems had not been pushed aside.

A GRIM FUTURE

At the turn of the twentieth century, almost two million chimpanzees roamed the earth. By the end of 1999, it was estimated that only 220,000 were left spread across 21 nations. To help save the chimps in Africa, Goodall established a number of chimp sanctuaries in several countries. In 1992, the Tchimpounga Sanctuary for Chimpanzees was established in the Congo Republic. In 1993, the Sweetwaters Sanctuary was organized in Kenya. Two years later, in 1995, the Kitwe Point Sanctuary for Chimpanzees was established in Tanzania. All of these havens offered a safe environment for chimps that were rescued from abusive situations, lost, orphaned, or ill. Goodall has also encouraged the opening of some areas to promote eco-tourism, where people can come and see the chimps, monkeys, and other animals in their native habitats.

As efforts to provide more safe areas for chimps increased, Gombe now faced growing threats from the outside. By the end of the twentieth century, it was becoming an isolated pocket of forest surrounded by bare hills and terrible soil erosion and dreadful poverty. As more people come to the area, the land is struggling to support the growing population. Chimps began disappearing, due to the destruction of the forest by the logging companies, or from snares and traps set for antelopes and pigs. An even grimmer picture presented itself; with a smaller population of chimps, inbreeding among surviving chimps weakens the gene pool. If it were to continue, the chimps at Gombe and other areas would vanish.

Despite the outside threats, Gombe continues to conduct the longest-running study of wild animal behavior in the world. While experts still come to the center to work with chimps, increasingly a good deal of the research is done by Tanzanians from surrounding villages. This has not only prevented poaching, but with programs such as TACARE, people are beginning to live better lives. For the most part, it has been a successful effort. The people who are part of the Gombe project know that Goodall and the center care about their well-being as well as that of the chimpanzees.

CARRYING THE MESSAGE

Since 1986, Goodall has traveled almost nonstop and has not stayed anywhere longer than three weeks. There are days when she sometimes wonders where she is. Often on her trips to the United States, Goodall rarely spends more than two nights in one place. There are always lectures to give, new people to meet, receptions, and press conferences. She spends countless hours on airplanes, writing letters, and sorting slides. Wherever she travels she carries brochures from the Jane Goodall Institute. She is often accompanied by someone from the Institute who shields her. Given her grueling schedule, employees of the Institute take on different legs of the trip; this way Goodall never is alone. She has trouble sleeping and admits that she has probably taken on too much. But when she closes her eyes, all she can see are the images of chimps in cages, small, orphaned youngsters, or chimps being abused, tortured, or killed. As she sadly described it, she thinks about tiny chimps in tiny prisons even though they have committed no crimes. And once anyone has seen it, they will never forget it.

In the hustle and bustle of her travels, Goodall still manages to get away for family time. When in England she stays at the Birches in Bournemouth where she grew up. In her old bedroom, her first chimpanzee, Jubilee rests on her bed. The pace of life is a little slower there; but her days are full. Breakfast comes at nine o'clock; by that time Goodall has already been at work for three hours. During that time she catches up with her correspondence and tries to answer as many letters as she can, particularly the ones from children. She works on her books and articles. In the afternoon she continues her writing, then stops to take tea with the family and walk her dog. She eats dinner, and then walks her dog once more. At night, she returns to work before turning in for the day.

Goodall remains close to her son, Grub, who is now married with two children. Grub still shows no interest in chimpanzees, but his son has shown great interest in his grandmother's work, and has the same passion for animals that Goodall showed as a young girl. There have been sad times, too. In 2000, her first husband, Hugo Van Lawick, died in Tanzania, having gone to live there with Grub and his family. Even more devastating was the passing of her mother, Vanne, in 2001. Goodall still speaks of her and the tremendous influence she had on Goodall's life. Vanne was more than a mother; she was Goodall's sounding board, traveling companion, and most of all her greatest supporter. Goodall acknowledges that without a mother like Vanne, she might never have achieved her dream of going to Africa and living with the chimpanzees.

40 YEARS AT GOMBE

As she moved into the twenty-first century, Goodall showed no signs of slowing down. At 70, she continues to travel almost 300 days a year, bringing her message of caution and hope. She enjoys an almost rock-star status as she winds her way around the globe. She has expanded her presence by appearing in numerous television programs, such as documentaries on HBO and Animal Planet. She is the subject of several biographies for children. Despite the early criticisms of Goodall by the feminist movement, there are lines of young girls who wait to catch a glimpse of her, or, if they are lucky, to speak with her. She is stopped constantly for autographs. She has become a hero to many, crossing the lines of age, gender, race, creed, and color.

Her organization and programs are powerful forces in promoting chimpanzee survival and welfare. Along with the CCCC, the Jane Goodall Institute has become a formidable lobby for animal rights. Besides the Chimpanzoo project and Roots & Shoots, the Gombe research center has also expanded its mission by providing teaching aids and information packets to teachers and schools all over the world. The center also helps educate tourists and visitors about chimpanzees.

Her honors also continue to accumulate. She has been recognized numerous times in various countries for her contributions. Her homeland has not forgotten her either; in 1995, Goodall was awarded the title CBE (Commander of the British Empire) by Queen Elizabeth II; in 2004, she was made a Dame of the British Empire, one of highest honors given to English women. Prince Charles, who conferred the award, is himself an ardent supporter of animals and the environment. Her attitude toward these and many other awards is self-deprecating and low-key. The awards are secondary; what she has done to earn them, whether it's raising funds or helping her advance her message, is what matters.

The United Nations recently appointed Goodall as one of a handful of celebrity "Messengers of Peace" to help publicize the mission and work of that organization. Goodall has taken her new role seriously, and has campaigned against the war in Iraq. The world's troubles still cause her despair from time to time. But she shakes it off and gets going again. She has no plans to take things easy. But people know that there will come a point when Jane Goodall will no longer be able to continue. At the Jane Goodall Institute, people have already started to plan on how they will continue Goodall's work without her. Indeed, many worry that life after Jane is gone might bring to a grinding halt all that she has accomplished over the last four decades.

AN ONGOING BATTLE

Goodall still has her share of detractors. There are those in other conservation and animal protection groups that resent her abilities to capitalize on her name. Many grumble that Goodall, because of her success, has taken the biggest piece of the funding and donations pie, while they get the crumbs. In truth, the Jane Goodall Foundation still struggles for money; by all estimates Goodall's speaking engagements generate upward of $150,000 a year. But with 3,000 employees, the CCCC, and Gombe, the funds are often stretched thin. In reality, Goodall is supporting three of her dearest children; she will never give up fundraising as long as she can carry on.

There are those, too, who have misunderstood Goodall's beliefs and views. A confirmed vegetarian, she does not condemn those who eat meat, but those who raise animals in an inhumane way. At one conference, she was confronted by a woman who decried her criticism of medical research, telling her that it was necessary to save lives. Goodall gently pointed out to the woman that she knows the importance of medical research; her second husband died of cancer, while her mother's life was saved because of a pig's heart valve. To stop medical research would harm countless individuals and that she does not intend. What Goodall protests is the idea that humans might stand to profit from medical advances against the welfare of animals.

Goodall knows that people who have benefited from animal's contributions are valuable allies. They owe their lives to animals and, if approached in the right way, will gladly make donations for the humane care of animals because of what the animals have done for them. When confronted by an official of the NIH (National Institute of Health) who told her that it would cost too much to institute improvements in the treatment of lab animals, Goodall asked him to think about his house, his office, his car, his administration building, and the holidays he took. She then asked him to tell her why he would want to deprive these animals of the care and compassion they so richly deserve. In discussing some of the controversies with reporters, she said: "The feeling has been expressed that if Jane gets her way with chimps, that's the thin end of the wedge. Next they say, she's going to want to improve conditions for monkeys, dogs, and all the other animals. You bet I do."[4]

A LABOR OF LOVE

Today Goodall's work has been accepted by researchers all over the world. The young woman who was considered a dumb blond and a pinup for the National Geographic Society is now considered unparalleled in her field. She no longer has to listen to "You Tarzan. Me Jane" jokes or put up with hurtful comments about her demeanor or dress. Where critics once charged that the only reason Goodall became popular was that she looked good in shorts, she has millions of supporters who see her as an oracle for the future and leader for the young.

Her writings today continue unabated; they, too, have taken many twists and turns. From her first article in 1963 for the *National Geographic*, which was so popular the issue sold out, to the monumental work *The Chimpanzees of Gombe*, Goodall offered a window into the world where she had lived and worked. Later, her writings took on a more spiritual tone as she contemplated her life and work in a series of autobiographical books. Goodall has never been ashamed of her faith, and has no problem reconciling it with science. For her, faith helped define her place and role in the universe; while some find it at odds with science, precision, and objectivity, Goodall sees no conflict with it all.

For Goodall, science has been about the small "s" rather than the large "S," which requires wearing a white coat and being objective and emotionally detached from your subject. In her years of studying the chimpanzees at Gombe, Goodall turned that idea on its head; instead of documenting statistics and objective observations she told stories. She gave her subjects names instead of numbers. She watched their personalities and their movements. Most important, she went into the field without any beliefs or theories that would guide her to a preconceived conclusion.

Studying and respecting the chimps was the most important key to her success. She was the trespasser, the interloper who had been allowed in the door. She often told her students that if they encountered a chimp on a trail coming the other way, they were to step aside and let the chimp pass. Remember, she said, they belong here, you do not.

GOODALL'S LEGACY

Jane Goodall is a good storyteller. It is one of the gifts that allowed her to convey her thoughts, hopes, and concern for her subjects. In the more than 40 years that she has studied chimpanzees she has brought them to life, whether through books, films, or television programs. She has helped show their lives in a three-dimensional way; they are no longer relegated as objects or toys, but are seen as breathing, living, emotive beings. Some of her chimps have attained an almost iconic place in our culture: one can now buy a poster of David Greybeard and note cards decorated with a picture of the Gombe forest. People can donate money to sponsor a chimp. One of the more popular items is Mr. H, a model of a plush monkey that Goodall received in 1996, and which she has taken with her around the world. Since then, Mr. H has been to over 50 countries and has been touched by 2 million people. He has become the Jubilee of a new generation of children.

In her work, Goodall made, and continues to make, important connections between man and chimp. She understands that as humans evolved physically from an ape-man, they have culturally continued to evolve; and that evolution needs to concentrate on developments that make humans more compassionate and less aggressive. She says that what humans evolved from is not nearly as critical as what they are in the process of becoming. There is an interdependence between man and everything in nature and that must be respected, taken great care of, and understood.

It was Goodall who discovered that chimps are more like ourselves than we had previously imagined. They wage wars over territories. They fashion crude weapons. They make and use tools to help them find food. Chimps exhibit real emotions: love, jealously, selfishness, anger, sadness, and grief. They embrace each other, they play with each other, and they take care of each other. Their society is ordered; there is continuity to their lives. They are parents, children, teachers, leaders, mediators, victims, and aggressors. They can be taught sign language but do not understand past or future meanings. Chimps are not capable of "discussing the distant past and learning from it."[5] But that is something that humans can do. By understanding the strong connection to the "mock man," humans can begin to answer the question Louis Leakey asked himself, "Who am I and where do I come from?"

Goodall misses the idyllic first years at Gombe. Then it was a paradise and refuge from the outside world. But Goodall realized that if something or somebody gives to you, you owe something in return. In 1960, Goodall was presented with an amazing opportunity. She took it and in doing so changed not only how we look at chimpanzees and other primates, but also how we learn about them. Her job now is to save what she loves. For her, the way to move forward is to get people's hearts involved. In that, she has been uncommonly successful. She still manages to return to Gombe whenever she can. Often her visits are spent walking the hills and watching the chimps. One of her greatest pleasures is observing mothers and children; seeing them at play or caring for their children offers hope that they and their world will not disappear. But she also remembers the past, sometimes with sadness, sometimes with joy. The ghosts of her first friends, David Greybeard, Flo, William, Mr. McGregor, and so many others do not haunt her; they guide her. In the end, Jane Goodall's life is not about fame or success. It is not about the quest for objectivity. It has always been, and will continue to be, about what she "owes to the chimps."

There is a story that Jane Goodall likes to tell when she speaks. There was a chimpanzee named Old Man who was not really old; he just looked old. Perhaps this was because of his earlier life. When he was two years old, living in Africa, his mother was killed and he was shipped to North America to spend the next 10 years of his life in a biomedical research lab. When he was 12, he was retired to a zoo where he was placed with two other females and a male. His caregiver was told to be careful, as Old Man hated people and would try to kill him.

But slowly, the caregiver built a relationship with Old Man. He gave the chimp bananas and groomed him. One day, the caretaker slipped, startling one of the chimpanzee babies. As the baby screamed, the mother chimp raced over and bit the caretaker's neck. Then two other chimpanzees ran to the mother's aid; one bit the caretaker's leg, the other his wrist. Suddenly Old Man came thundering toward the caretaker, his hair bristling in rage. The caretaker thought Old Man too would also attack him. Instead, Old Man pulled the other chimpanzees off the caretaker and kept them away while the man dragged himself to safety. After a pause, Goodall then asks her audience, "If a chimpanzee can reach across the species gap to can save a man, surely we can reach out to help the chimpanzee."[6]

NOTES

1. Nick Schoon, "Jungle Girl Grows Up," *Saga Magazine* (April 2004) http://www.saga.co.uk/magazine/pages/article.asp?id=DF3C2D2-9B12-47E9-9477-926113E5EBAB&bhcp=1 (accessed October 1, 2004).

2. Sy Montgomery, *Walking with the Great Apes: Jane Goodall, Dian Fossey, Biruté Galdikas* (Boston: Houghton Mifflin, 1991), p. 200.

3. Sy Montgomery, *Walking with the Great Apes: Jane Goodall, Dian Fossey, Biruté Galdikas* (Boston: Houghton Mifflin, 1991), p. 199.

4. Sy Montgomery, *Walking with the Great Apes: Jane Goodall, Dian Fossey, Biruté Galdikas* (Boston: Houghton Mifflin, 1991), p. 211.

5. Alan Alda, "Frontiers Profile: Jane Goodall," PBS-Scientific American Frontiers, Chimps R Us, http://www.pbs.org/saf/1108/features/goodall3.htm (accessed September 10, 2004).

6. "Jane Goodall," *Current Biography Yearbook, 1991* (New York: H. W. Wilson Company, 1992), p. 253.

BIBLIOGRAPHY

Addley, Esther. "The Ascent of One Woman." *Guardian Online*, http://education.guardian.co.uk/academicexperts/story/0,1392,928205,00.html (accessed June 1, 2004).

African Wildlife Fund. "Vervet Monkey." http://www.awf.org/wildlives/154 (accessed August 5, 2004).

Alda, Alan. "Frontiers Profile: Jane Goodall." PBS–Scientific American Frontiers, Chimps R Us, http://www.pbs.org/saf/1108/features/goodall3.htm (accessed September 10, 2004).

American Psychology Association. "APA Historical Database: Selected Entries." http://www.cwu.edu/~warren/calendar/cal0205.html (accessed July 1, 2004).

Arias, Ron. "Jane Goodall." *People Weekly*, May 14, 1990, pp. 93–96.

"Chimpanzee Vocal Communication." Discover Chimpanzees, http://www.discoverchimpanzees.org/activities/pant_hoots.php (accessed July 1, 2004).

"The Chronicle Interview." *U.N. Chronicle*. (Sept.-Nov. 2002): 25+.

Dale, Steve. "An Interview With Jane Goodall: What I Learned From Dogs." Studio One Networks, http://www.thedogdaily.com/you_dog/moments/archive/goodall_interview/ (accessed November 15, 2004).

"Discover the Serengeti." http://www.serengeti.org/ (accessed November 1, 2004).

Goodall, Jane. *Africa in My Blood: An Autobiography in Letters, The Early Years*. Ed. Dale Peterson. New York: Houghton Mifflin, 2000.

———. *Beyond Innocence: An Autobiography in Letters, The Later Years*. Ed. Dale Peterson. New York: Houghton Mifflin, 2001.

———. *The Chimpanzees of Gombe: Patterns of Behavior*. Cambridge MA: University of Harvard Press, 1986.

———. *40 Years at Gombe*. New York: Stewart, Tabori & Chang, 1999.

———. *In the Shadow of Man*. Boston: Houghton Mifflin, 1971.

———. "Life and Death at Gombe." *National Geographic* (June 1979): 592–621.

———. My *Friends the Wild Chimpanzees*. Washington, D.C.: National Geographic Society, 1967.

———. My *Life with Chimpanzees*. New York: Simon and Schuster, 1996.

———. *Through a Window: 30 Years Observing the Gombe Chimpanzees*. London: Weidenfeld & Nicolson, 1987.

Goodall, Jane, with Marc Bekoff. *The Ten Trusts: What We Must Do to Care for the Animals We Love*. San Francisco: Harper, 2002.

Goodall, Jane, with Phillip Berman. *Reason for Hope: A Spiritual Journey*. New York: Warner, 1999.

Goodall, Jane, and Michael Nichols. *Brutal Kinship*. New York: Aperture Foundation, 1999.

Goodall, Jane, and Dale Peterson. *Visions of Caliban*. Boston: Houghton Mifflin, 1993.

Goodall, Jane, and H. Van Lawick. *Innocent Killers*. Boston: Houghton Mifflin; London: Collins, 1970.

"History of Physical Anthropology." http://uts.cc.utexas.edu/~bramblet/ant301/two.html#anchor647252 (accessed August, 2004).

Jahme, Carol. *Beauty and the Beasts: Woman, Ape and Evolution*. London: Virago, 2000.

"Jane Goodall." The Jane Goodall Institute, http://www.janegoodall.ca/jane/jane_bio_early.html (accessed October 10, 2004).

"Jane Goodall." *Current Biography Yearbook, 1991*. New York: H.W. Wilson Company, 1992, pp. 249–253.

"Jane Goodall." *Newsmakers 1991*, Issue Cumulation. Gale Research, 1991. Reproduced in *Biography Resource Center*. Farmington Hills, MN: Thomson Gale, 2004, http://galenet.galegroup.com/servlet/BioRC (accessed, May 5, 2004).

"Jane Goodall." *New York Times Magazine*, February 18, 1973, pp 14+.

The Jane Goodall Institute. "Curriculum Vitae." http://www.janegoodall.org/jane/cv.html (accessed June 1, 2004).

Kohler, Wolfgang. *The Mentality of Apes*. London: Routledge and Kegan Paul, 1925.

Lassieur, Allison. "When I Was a Kid: Childhood Experiences of Famous People." *National Geographic World* (September 1999).

Lee, Luaine. "Return to Gombe with Jane Goodall on Animal Planet." *Knight Ridder/Tribune News Service*, Feb 23, 2004.

Lindsey, Jennifer, and the Jane Goodall Institute. *Jane Goodall: 40 Years At Gombe*. New York: Stewart, Tabor & Chang, 1999.

Mapes, Jennifer. "Jane Goodall: 40 Years in Africa." *National Geographic News* (March 13, 2001), http://news.nationalgeographic.com/news/2001/03/0313_goodall.html (accessed September 1, 2004).

McCarthy, Susan. "Jane Goodall: The Hopeful Messenger." October 27, 1999, http://www.salon.com/people/feature/1999/10/27/reason (accessed October 15, 2004).

Michaels, Marguerite. "Oct. 20, 1952: The Bloody Mau Mau Revolt." (Special Section, *Time*'s 80th Anniversary/80 Days that Changed the World) *Time*, March 31, 2003.

Montgomery, Sy. *Walking with the Great Apes: Jane Goodall, Dian Fossey, Biruté Galdikas*. Boston: Houghton Mifflin, 1991.

Morell, Virginia. *Ancestral Passions: The Leakey Family and the Quest for Humankind's Beginnings*. New York: Simon & Schuster, 1995.

———. "Lost Chimps: Primatologist Jane Goodall Has Taken on a New Role as a Savior to Africa's Orphaned Chimpanzees." *International Wildlife* (Sept-Oct 1996): 12–22.

Morris, D., and R. Morris. *Men and the Apes*. New York: McGraw-Hill, 1966.

NationalGeographic.com, "Explorers-in-Residence: Jane Goodall." December 7, 2001, http://www.nationalgeographic.com/eir/bio_goodall.html (accessed June 30, 2004).

Nissen, H.W. "A Field Study of the Chimpanzee." *Comparative Psychology Monograph*, 8, (1931): 1–22.

"Obituary." *London Sunday Times*, September 1972.

The Public Broadcasting Corporation. "Jane Goodall's Story." http://www.pbs.org/wnet/nature/goodall/story.html (accessed July 1, 2004).

"Robert Mearns Yerkes." Human Intelligence, http://www.indiana.edu/~intell/yerkes.shtml (accessed June 10, 2004).

Salon.com. "People: A Conversation with Jane Goodall." December 7, 2001, http://www.salon.com/people/feature/1999/10/27/goodallint/ (accessed June 30, 2004).

Schoon, Nick. "Jungle Girl Grows Up." *Saga Magazine*, April 2004, http://www.saga.co.uk/magazine/pages/article.asp?id=DF3C2D2-9B12-47E9-9477-926113E5EBAB&bhcp=1 (accessed October 1, 2004).

Scientific American. "Profile: Jane Goodall." December 7, 2001, http://www.sciam.com/1097issue/1097profile.html (accessed July 1, 2004).

"Some Important Figures in Primate Issues and Research." http://pubpages.unh.edu/~jel/512/primate_people.html (accessed October 1, 2004).

Toms, Michael. "Born To Be Wild: A Conversation with Jane Goodall." New Dimensions, World Broadcasting Network, 1999, http://www.newdimensions.org/online-journal/articles/born-to-be-wild.html (accessed June 10, 2004).

"Years of Watching Chimps Helps Us Understand Ourselves." (A Conversation With Jane Goodall) *U.S. News & World Report,* Nov. 5, 1984, p. 81.

Yerkes, R.M. *Chimpanzees: A Laboratory Colony.* New Haven, CT: Yale University Press, 1943.

Yerkes, R.M., and A.W. Yerkes. *The Great Apes: A Study of Anthropoid Life.* (New Haven, CT: Yale University Press, 1929).

WEB SITES

Chimps R Us http://www.pbs.org/saf/1108/index.html

Jane Goodall Center for Primate Studies http://www.discoverchimpanzees.org/

The Jane Goodall Institute http://www.janegoodall.org/

Jane Goodall's Wild Chimpanzees http://www.pbs.org/wnet/nature/goodall/index.html

Roots & Shoots http://www.rootsandshoots.org/

INDEX